Climate Change Impact on Ecosystem

About the Editors

J. Sundaresan, Head, Climate Change Informatics in CSIR- NISCAIR an autonomous institution under Council of Scientific and Industrial Research (CSIR), Ministry of Science & Technology, Govt of India. He is the Editor of Indian Journal of Geo-Marine Sciences (IJMS), a publication of CSIR. He is a PhD in Oceanography from Cochin University of Science & Technology, India. Thirty years research experience in coastal oceanography and twenty two years experience in Island studies. Associated with science education, science popularisation and ecological activities for the last 36 years. Actively associated with scientists Associations to sustain ethics, truth and scientific temper in scientific institutions. Published more than fifty publications, an examiner for doctoral thesis evaluation and member of national and international committees. Invited for key note address and invited talk in many Universities. Travelled in many countries.

S. Sreekesh has about 18 years of experience in teaching and research. He obtained his Ph D degree in land degradation in Periyar river basin from the Jawaharlal Nehru University, New Delhi and worked with The Energy and Resources Institute and the TERI University, New Delhi. He is currently working as an Associate Professor in the Centre for the Study of Regional Development, Jawaharlal Nehru University, New Delhi, India. He is actively engaged in teaching and research in the field of climate, climate change and water resources management. He has expertise in application of geospatial techniques in resource and environment management. He has published many research papers in the national and international journals and wrote several chapters in books. He also has vast experience in collaborative research with the support from the national and international funding agencies.

AL. Ramanathan is a Professor in School of Environmental sciences, Jawaharlal Nehru University, New Delhi, India. He is leading the group working on coastal biogeochemistry and hydro geochemistry and has worked extensively on mangroves estuaries and coastal ground waters of India for the past two decades. He is actively engaged in coastal research with different institutions in India and from various parts of the world such as Australia, Russia, USA, etc, on nutrient dynamics, Biomarkers, paleo environment, nutrient source identification etc., He has guided two dozen PhDs in these aspects and published more than seven dozen papers in referred reputed journals like Estuaries, Wetland Ecology and Management, Estuarine, Coastal and Shelf Science, Journal of coastal research, Bull of Marine sciences, Indian journal of marine sciences, Marine pollution bulletin, Hydrogeology Journal, Water Soil and Air Pollution, Hydrological Process , Geofluids Applied Geochemistry, Journal of Geochemical Exploration etc., He has also published five books and wrote several chapters in books of national and international repute. He was a Post doc fellow under STA Japan program,, UGC Russian academy of science program, CSIR, INSA, DST, etc., He was a member of editorial board in Indian journal of marine sciences and is a referee for many national and international journals.

Leonard Sonnenschein, 46 years experience in keeping fish, 35 years experience in scientific research, 23 years experience in science education innovation, over 100 publications, and extensive performance in conservation collaboration, climate change issues and public awareness. Leonard Sonnenschein opened St. Louis Children's Aquarium in 1993 and on June 8, 2004 (World Ocean Day) opened its expansion facility, the World Aquarium. Leonard regularly supervises students from over 45 universities which collaborate with the research component of the aquarium in facility development, exhibit design, fisheries, aquatic sciences, ecology, aquariology, legal frameworks, consumer awareness, cultural comprehension of environmental issues, and public understanding through field, conference and inter-governmental work. In 2006, Sonnenschein founded the Conservation for the Oceans Foundation to expand the World Aquarium's focus. The mission of the Conservation for the Oceans Foundation (CFTO) is to support grassroots-level conservation, education and research projects that bring about positive changes to ecosystems worldwide through local and multi-stakeholder actions. In 2009, Youth Voices in Conservation was developed for additional youth engagement opportunities (ages 3-50). In 2011, based upon the Low Carbon Lifestyles' campaign, the Youth Voices in Conservation's GreenLeaf Program was developed to allow for raising capital for residual support mechanism based upon the carbon offset credit from worldwide projects' actions. Leonard regularly collaborates with international agencies such as UNESCO, UNEP, WHO, International Ocean Institute, the Global Forum on Ocean, Coasts and Islands and is a co-founder of the World Ocean Network. Leonard recently started Innovative Drug Manufacturing, LLC to bring new patented technology to the pharmaceutical, nutraceutical, cosmetic and aquaculture industries.

Ram Boojh, Programme specialist at the UNESCO Office in New Delhi, responsible for Ecological and Earth Sciences, World Heritage Biodiversity and natural heritage programmes and is also focal point for the UN Decade for Education for Sustainable Development (DESD). He has over 30 years of experience of working with the academic institutions, voluntary sector, Government and international organizations. He has a distinguished academic career with Doctorate in Ecology and recipient of many awards and honours including the Indian National Science Academy Medal presented by Mrs Indira Gandhi, then Prime Minister of India in January 1984. He has travelled widely and has been visiting fellow at many European and US universities and academic institutions. He has Published over 100 research/technical papers / popular articles and 11 books.

Climate Change Impact on Ecosystem

Editors

J. Sundaresan
S. Sreekesh
A.L. Ramanathan
Leonard Sonnenschein
Ram Boojh

SCIENTIFIC PUBLISHERS (INDIA)

Published by:

Scientific Publishers (India)
5 A, New Pali Road, P.O. Box 91
Jodhpur 342 001 (India)

Branch Office
Scientific Publishers (India)
4806/24, Ansari Road, Daryaganj
New Delhi - 110 002 (India)

E-mail: info@scientificpub.com
Website: www.scientificpub.com

ISBN: 978-81-7233-832-9

Typeset: Rajesh Ojha
Printed in India

प्रो. समीर के. ब्रह्मचारी
महानिदेशक, वै.औ.अं.प.
एवं सचिव, भारत सरकार
वैज्ञानिक तथा औद्योगिक अनुसंधान विभाग

Prof. Samir K. Brahmachari
Director General, CSIR
& Secretary, Government of India
Department of Scientific & Industrial Research

वैज्ञानिक तथा औद्योगिक अनुसंधान परिषद्
अनुसंधान भवन, 2, रफी मार्ग, नई दिल्ली – 110 001
COUNCIL OF SCIENTIFIC & INDUSTRIAL RESEARCH
Anusandhan Bhawan, 2, Rafi Marg, New Delhi - 110 001

Foreword

Climate, change, and concomitantly the deteriorating environment, is a huge concern world-wide, with scientific, social and economic dimensions. Climate Change is a conspicuous problem and poses immense challenges to a secure and civilized society on planet earth. The World Health Organisation (WHO) estimates that the warming and precipitation trends due to anthropogenic climate change of the past 30 years is already claiming over 150,000 lives annually. Many local and global issues prevalent in human society today, from crop failures to diseases like respiratory illnesses due to heat wavs, and altered transmission of infectious diseases are linked to climate fluctuations. Evidence of changes in geological, physical and biological systems is being observed globally. Intergovernmental Panel on Climate Change (IPCC) has constituted many working groups to analyze evidences form various such systems. Studies undertaken by these groups invariably suggest that almost all the natural systems are being affected by regional climate change especially the rise in temperature. The impacts of regional climate changes are reflected primarily in the prevalent terrestrial, biological and hydrological systems. This has resulted in significant and often drastic changes in agriculture, forestry, surface and ground water availability. However, climatic forcing factors that affect the natural systems are yet to be fully analyzed and understood. The changes in climate can be attributed to natural and anthropogenic reasons or may also be due to a collective attribution of both factors. It is much needed to initiate joint ventures to examine the basic reasons of regional climate change. We also need to identify and develop reliable indicators of regional climate change. It is indeed the need of the hour to explore, discover and extend novel strategies to both, adapt to the changes and to mitigate its effects.

Though responses on regional climate change caused due to anthropogenic reasons are specific in few continents, data and published studies, pertaining to examination of anthropogenic climate forcing factors in many of the continents are by far very few. Studies and databases on climate forcing factors and parameters are relatively scarce in South Asia, Africa, South America and Indian Ocean Regions. Improved observations and networking are inevitable to these regions.

Present Book, Climate Impact on Ecosystem in conspicuous due to its spatially and geographically distributed studies on impact of climate change on multi-system environment i.e. physical, geographical and biological systems, in the data scarce regions of Indian subcontinent. this book includes studies on the impact of climate change on water resources of islands in Gulf of Mannar to '*Jumping*' a traditional farming method adopted by tribal people residing in the North Eastern hills of Himalayas. This book throws light on the various methods that are being developed and adopted to combat the adverse effects of climate change in States like Assam and discuss about the presently available innovative facilities like the high efficiency traditional chullas, designed to reduce carbon emission and thus prevent climate change. Multidisciplinary innovations and initiatives are inevitable to gain a deeper insight on the impact of climate change on ecosystem. This requires, high level computational modeling and simulation. This book gives an insight on the use of computational techniques to model the impact of temperature change on environment. Though governments of each nation is taking efforts to identify the implement adaptive measures to mitigate the effects of climate change a joint global venture in this regard is absent. This book hence deliberates on the lack and need of an effective international agreement to identify measures to mitigate climate change.

This book tries to explore and explain the prevasive effects of climate change on ecosystem and tries to conceptualize an integrative approach in place of sectoral traditional management system. The book is a good compilation of articles focusing on issues pertaining to difficult types of ecosystem and I extend the authors and contributors my appreciation for such a rare collection on climate change impact studies. I wish this book large circulation and I am sure that this will prove to be high beneficial to researchers, policy makers and administrators.

[Samir K Brahmachari]

New Delhi
October 31, 2012

Preface

Climate change is a fact. It is evident from the physical, biological and geological system responses from all the continents and oceans. According to various working Group of IPCC (Intergovernmental Panel on Climate Change) anthropogenic outcome had significantly contributed to the global warming and climate change. Evidence of global and continental changes due to global warming are conspicuous in continents like Europe, Central Asia and North America where the databases and studies related to different environmental parameters are noteworthy. Databases and published information related to climate change responses are sparse in regions like Southeast Asia, Indian Ocean and Africa Present book consists studies on climate change from Indian continent. This book "climate change impact on ecosystems" consists observations from different eco systems.

Regulatory measures for judicious utilization of natural resources are essential. The first chapter of this book discusses the environmental impact assessment and trend and practices for sustainable development. Green house gas emissions are to be reduced so as to minimize the climate change and process. Possible ways to control the factors that affect climate change are discussed in the second chapter. It suggest many recommendations to improve the environmental scenario of Assam the foothills of eastern Himalaya.

Fresh water contamination due to saline water intrusion in coastal and Island aquifers is the immediate effect of sea level rise due to climate change. A case study was carried out in Kurusadai Island, Gulf of Mannar for adequate water availability is ascribed in this book . Rain water may drain through surface and subsurface into ocean, soon after the rain. The number of low pressure system formed over Indian landmass and adjoin seas are examined in this book. Low pressure system shows fluctuating trend in Arabian Sea and Bay of Bengal. Hundred years data on temperature and rainfall were analyzed in another study It is observed that during the period of high global temperature, rainfall occurrence is low.

Sources of climate change induced gases are presented in two chapters. Environmental issues related to open mining may augment the process of global warming. Chemistry of lignite and coal consuming power plants and fly ash are examined. Coal power plant through exit hot flue gas has substantial contribution towards climate change. Agriculture is one of the source of green house gases that ultimately increase regional and global temperature. Jhuming, a traditional method of farming in Manipur a hilly mountain state is discussed in this book. Jhuming is directly related to deforestation and destruction of biodiversity. Socio-economic and ecological impact of Jhuming is also presented.

Climatic zones of different geographical locations are an important aspect of Climate change. Brown seaweeds from Karachi coast is examined in this book. Ecosystem and eco subsystem parameters of major and minor estuaries in India reveal the oxygen deficiency in surface and subsurface waters of many estuaries. A typical mass balance model was developed for dissolved oxygen for Krishna estuary. About 100 botanic gardens located in different parts of Indian subcontinent are placed in network as a part of the adaptive measures to climate change. This book also focused on the availability of Rare, Endangered and Threatened (RET) species in Indian Botanic Gardens. There are as many as 565 RET species are growing in 33 botanic gardens of India.

Models developed on the basis of Neuro-Fuzzy technique is discussed in this book for regional climatic variation. After model development, the effects of rise in temperature (Global Warming) have been studied over potential evapotranspiration. This will presents the climate change scenario in this region. Remote sensing technology can be an effective toll to examine the same. This book consists studies related to Ogni storm that crossed Indian coast on October 30, 2006. IRSP6LISS image was utilised for the study. Envisat satellite were used for submergence analysis due to Ogni storm.

Impact of climate change in Indian economy and environment are discussed in this book. India should modulate an effective international agreement to deal with issues related to Climate Change. Weather warning can be disseminated to the stakeholders by way of improved Information and communication technology. A new model on dissemination of communication of weather forecast in line with the global standards meeting the Millennium Declaration Target is discussed in this book.

Many of the activities viz. use of LPG, use of waste paper by millions of families and water conservation have significant contribution to global warming. Household behavior is the nerve centre for mitigating global warming. Efficiency of traditional chullas will reduce the quantity of CO_2 emission to the atmosphere. Enhancing the efficiency one third of the fuel used by the traditional chullas may be saved. This book consists many devises developed and applied to reduce the carbon emission during domestic firewood use.

Climate change impact database and information are quite sparse in tropical regions. Present book consists multidisciplinary studies from different geographical and ecosystems of Indian subcontinent. The information available in this book varies from impact of climate change on mountain to Island aquifers. Studies on mines, industry, agriculture, coal and jhuming had made this book, a unique compilation of climate change information from a tropical country. It consists many tools related to mitigation especially to reduce the green house gas emissions from firewood during traditional cooking. The innovations and the creativity of the studies in this book will be highly beneficial for the future climate proof planning of tropical regions.

This book is a contribution of dedicated scientists and researchers who met at Cochin University of Science and Technology and Kadmat Island of Lakshadweep to discuss the impact of Climate Change on different ecosystems. Excellent support secured by School of Marine Sciences, Cochin University of Science Technology, Administrator U.T of Lakshadweep and Officers of Lakshadweep Science & Technology U.T of Lakshadweep, India are gratefully acknowledged.

J Sundaresan
S.Sreekesh
A.L Ramanathan
Leonard Sonnenschein
Ram Boojh

Contents

Environmental impact assessment — Trends & practices for sustainable development

J. Ansari

Environmental Technology, CSIR-National Institute for Interdisciplinary Science & Technology Industrial Estate, Thiruvananthapuram - 695 019, Kerala, India
[E-mail: kollamansari@yahoo.com]

Introduction

The traditional decision-making process of project management often tended to maximize short term gains at the expense of incurring long-term environmental liabilities. In project/program planning, alternatives were often evaluated and considered purely to economic reasons, without afterthought as to the long-term consequences of selecting the cheapest but environmentally unacceptable alternatives. It has become a mandate to plan development in a manner that would make the best use of resources and minimize the adverse impacts of human actions on the environment and it is imperative to evolve a planning tool which could effect environmentally sound development. Environmental Impact Assessment (EIA) is one such a tool[1].

EIA is generally defined as a process which consists of identifying and predicting the impacts of a proposed project or action on the biogeophysical environment and on man's health and well-being and interpreting and communicating information about such impacts in a manner which can be utilized by planners and decision makers.

EIA has been adopted and implemented by many developed and developing countries, after its introduction in the United States in the late 1960s. Positive benefits and significant improvements to planning design and decision-making resulting from EIA are widely recognized by many practitioners, organizations, and governments.

In Asia, EIA has been in practice for at least 25 years, but despite this, it has been relatively ineffective in protecting natural resources. The current development, industrialization and urbanization trends in the Asia region have culminated in extensive degradation of natural ecosystems and habitats leading to losses in biodiversity and unacceptable levels of urban pollution. In most of the region's cities, numerous pollution sources have resulted in severe contamination of water, air and soil. Untreated waste, sewage and uncontrolled traffic pollution are also contributing to deteriorating environments. Many of these problems are associated with a lack of adequate planning, poor institutional capacity, technical inadequacies, but above all, an apparent lack of awareness, concern and regard for these degraded environmental conditions. Major development projects relating to infrastructure, industrial and large power plant installations are often insensitively located, designed and managed resulting in severe interface problems with residential and community facilities.

Environmental protection and sustainable development are the milestones of the policies and procedures governing the industrial and other developmental activities in India. Ministry of Environment & Forests has taken several policy initiatives and enacted environmental and pollution control legislations to prevent indiscriminate exploitation of natural resources and to promote integration of environmental concerns in developmental projects. One such initiative is the Notification on Environmental Impact Assessment (EIA) of developmental projects issued on 27.1.1994. Till January 1994, obtaining environmental clearance from the Central Ministry was only an administrative requirement intended for mega projects. This notification includes details of procedures for obtaining environmental clearance and for public involvement.

In exercise of the powers conferred under the EPA-1986 the Union Cabinet on 18th May has accorded a new amendment for according prior environmental clearance under the National Environmental Policy. The background for Re-Engineering are:

- MOEF conducted a comprehensive review of the then EC process under the Environmental Management Capacity Building Project in 2001.
- Govindarajan committee set up by the Central Government for Reforming Investment Approvals and Implementation Procedures
- MOEF studies brought out the need for reforms, which are consistent with the Govindarajan Committee recommendations.

Table 1. Comparison of previous and existing systems

Steps	Previous system	How addressed in the present system
Screening	Related to project investment, and nature of activity without reference to likely impacts	Criteria are: Scale of impact, severity of impact, nature of location
Scoping	None: Project consultant has to guess at concerns of regulator	EAC specifies TORs
Public Consultation	Unstructured and duration not Under control of proponent	Clear structure, 45 days for PH (alternate provided if not done)
Appraisal	Centralized, uncertain with respect to time and information for decision-making	Decentralized. Defined time (60+15)/ frame

The Central Government hereby directs that on and from the 14th September, 2006[2] (date of publication) the required construction of new projects or activities or the expansion or modernization of existing projects or activities listed in the Schedule to this notification entailing capacity addition with change in process and or technology shall be undertaken in any part of India only after the prior environmental clearance from the Central Government or as the case may be, by the State Level Environment Impact Assessment Authority, duly constituted by the Central Government under sub-section (3) of section 3 of the said Act, in accordance with the procedure specified hereinafter in this notification (Table 1).

Categorization of projects

All projects and activities are broadly categorized into two categories – Category A and Category B, based on the spatial extent of potential impacts and potential impacts on human health and natural and man made resources.

All Category 'A' project shall require prior environmental clearance from the Central Government in the Ministry of Environment and Forests (MOEF).Category B can be B1 or B2 depending on the potential impacts. B projects to be appraised by State Environmental Impact Assessment Agency. In the absence of a duly constituted SEIAA or SEAC, a Category 'B' project shall be treated as a Category 'A' project.

Categorization of Projects

No.	Project	Category	Criteria
1.	Mining of Minerals	A	> 50Ha
		A/B	15-50 Ha
		B	< 50Ha
2.	Mineral beneficiation	A	> 0.1 mta
		B	≤ 0.1 mta
3.	Offshore/Onshore Oil & Gas exploration	—	All Projects
4.	River valley projects	A	> 20 MW/10,000 Ha
		B	< 20 MW/10,000 Ha
5.	Thermal Power Plants	A	> 500 MW (Coal/Gas), 50 MW (Diesel)
		B	< 500 MW (Coal/Gas), 50 MW (Diesel)
6	Nuclear Power	—	all projects
7	Cement Plant	A	≥ 1.0 mta
		B	< 1.0 mta
8	Petroleum Refinery	—	all projects
9.	Asbestos	—	all projects
10.	Chemical Fertilizer	—	all projects
11.	Pesticides	—	all projects
12.	Oil & Gas Pipeline	A	Project passing through more than 1 state
		B	All projects within a state

13.	Airports	—	all projects
14.	Ship Breaking Unit	—	all projects
15.	Industrial Estate, SEZ	A	≥50 Ha
		B	< 50 Ha
16.	Ports & Harbour	A	≥ 5 MTPA
		B	< 5 MTPA
17.	Roads	A	All new NH, Express highways
		B	All State Highway ≥ 30 km

Application for prior environmental clearance

An application seeking prior environmental clearance in all cases shall be made in the prescribed Form 1 and Supplementary Form 1A, if applicable after the identification of prospective site(s) for the project and/or activities to which the application relates, before commencing any construction activity, or preparation of land, at the site by the applicant. The applicant shall furnish, along with the application, a copy of the pre-feasibility project report except that, in case of construction projects or activities in addition to Form 1 and the Supplementary Form 1A, a copy of the conceptual plan shall be provided, instead of the pre-feasibility report.

Stages in the Prior Environmental Clearance (EC):-

- Stage (1) Screening (Only for Category 'B' projects and activities)
- Stage (2) Scoping
- Stage (3) Public Consultation
- Stage (4) Appraisal

Screening

In case of Category 'B' projects or activities, this stage will entail the scrutiny of an application seeking prior environmental clearance made in Form 1 by the concerned State level Expert Appraisal Committee (SEAC) for determining whether or not the project or activity requires further environmental studies for preparation of an Environmental Impact Assessment (EIA) for its appraisal prior to the grant of environmental clearance depending up on the nature and location specificity of the project. The projects requiring an Environmental Impact Assessment report shall be termed Category 'B1' and remaining projects shall be termed Category 'B2' and will not require an Environment Impact Assessment report. For categorization of projects into B1 or B2, the Ministry of Environment and Forests shall issue appropriate guidelines from time to time. The category also depends if they are proposed in notified ecologically/ sensitive areas like:

- No screening required for Category A projects
- Category B projects will be further screened at the state level for categorization into either B1 or B2. Specific guidelines to be evolved by MoEF

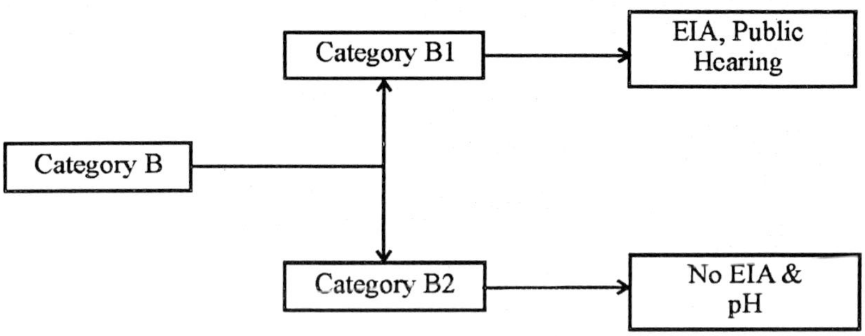

- Religious and historic places
- Archaeological monuments
- Scenic areas
- Hill resorts
- Beach resorts
- Coastal areas rich in mangroves, corals, breeding grounds of specific species
- Estuaries
- Gulf areas
- Biosphere reserves
- National parks and sanctuaries
- National lakes and swamps
- Seismic zones
- Tribal settlements
- Areas of scientific and geological interest
- Defence installations, specially those of security importance and sensitive to pollution
- Border areas (international)
- Airports

Scoping

"Scoping": refers to the process by which the Expert Appraisal Committee in the case of Category 'A' projects or activities, and State level Expert Appraisal Committee in the case of Category 'B1' projects or activities, including applications for expansion and/or modernization and/or change in product mix of existing projects or activities, determine detailed and comprehensive Terms of Reference (TOR) addressing all relevant environmental concerns for the preparation of an Environment Impact Assessment (EIA).

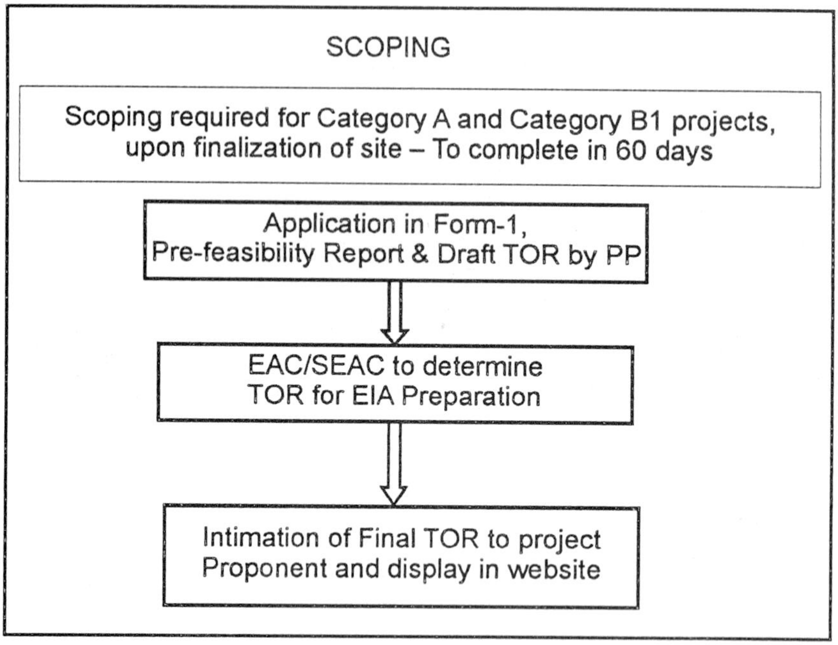

Report in respect of the project or activity for which prior environmental clearance is sought.

The Terms of Reference (TOR) shall be conveyed to the applicant by the Expert Appraisal Committee or State Level Expert Appraisal Committee as concerned within sixty days of the receipt of Form 1. If the Terms of Reference are not finalized and conveyed to the applicant within sixty days of the receipt of Form 1, the Terms of Reference suggested by the applicant shall be deemed as the final Terms of Reference approved for the EIA studies. The approved Terms of Reference shall be displayed on the website of the Ministry of Environment and Forests and the concerned State Level Environment Impact Assessment Authority.

Applications for prior environmental clearance may be rejected by the regulatory authority concerned on the recommendation of the EAC or SEAC concerned at this stage itself. In case of such rejection, the decision together with reasons for the same shall be communicated to the applicant in writing within sixty days of the receipt of the application.

Public consultation

This process helps to identify the concerns of local affected persons and others who have plausible stake in the environmental impacts of the project or activity.

All Category 'A' and Category B1 projects or activities shall undertake Public Consultation, except the following[4]:

- Modernization of irrigation projects.
- All projects or activities located within industrial estates or parks

- Expansion of roads and highways without acquisition of land.
- All Building/Construction projects/Area development projects and townships
- All category 'B2' projects and activities.
- All projects or activities concerning national defense and security.

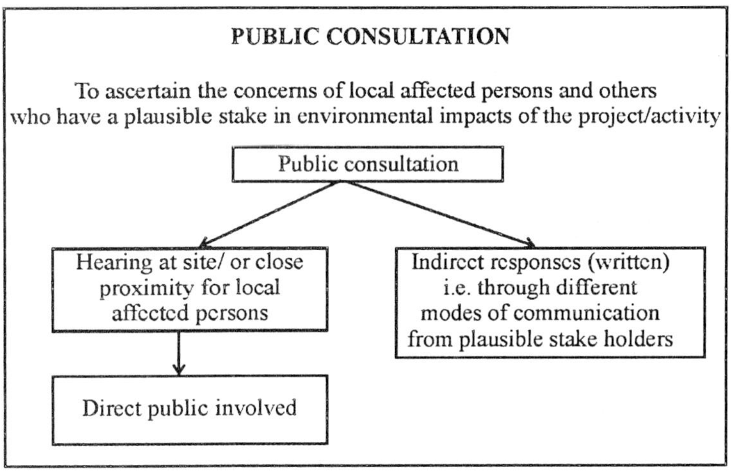

The public consultation is either conducted at site or at suitable place close to the site in the district. The District Collector is the chairperson of the public hearing committee. Other members of the committee includes the official from the district development body, SPCB, Department of Environment and Forest, Taluka and Gram Panchayat representative, and senior citizen of the district[3], etc. The hearing committee hears the objections/suggestions from the public and after inserting certain clauses it is passed on to the next stage of approval (Ministry of Forest and Environment). This process to be completed in 45 days of request by the project proponent.

In case the State Pollution Control Board or the Union Territory Pollution Control Committee concerned does not undertake and complete the public hearing within the specified period, and/or does not convey the proceedings of the public hearing within the prescribed period, the regulatory authority shall engage another public agency or authority which is not subordinate to the regulatory authority, to complete the process.

After completion of the public consultation, the applicant shall address all the material environmental concerns expressed during this process, and submits the EIA report with appropriate modifications. The final EIA report shall be submitted to the regulatory authority for onward process.

Appraisal

Appraisal means the detailed scrutiny by the Expert Appraisal Committee or State Level Expert Appraisal Committee considering the Final EIA report, report of public consultations. On completion of the process, the Expert Appraisal Committee or State Level Expert Appraisal Committee concerned shall make categorical recommendations to the regulatory authority concerned either for grant o prior environmental clearance on stipulated terms and conditions, or rejection o the application for prior environmental clearance, together with reasons for th same. The appraisal of an application shall be completed by the Expert Apprais Committee or State Level Expert Appraisal Committee concerned within sixt days of the receipt of the final Environment Impact and shall be placed before th competent authority for a final decision within the next fifteen days.

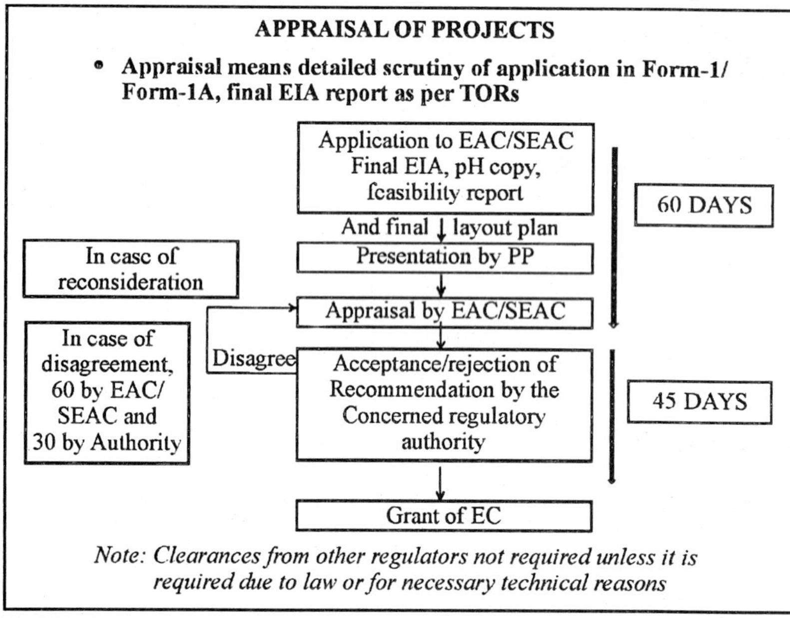

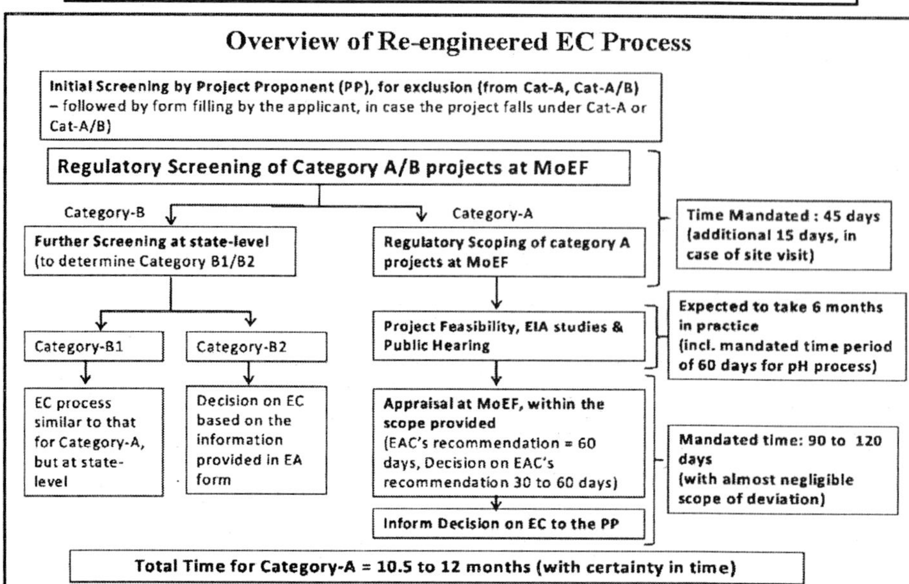

alidity of environmental clearance (EC)

he prior environmental clearance granted for a project or activity shall be valid or a period of ten years in the case of river valley projects, thirty years for ining projects and five years in the case of all other projects and activities.

IA as applied in mining project – A case study

uring initial screening, the specimen study area was classified into Category A. or scoping the project proponent submits the form 1 detailing the TOR of the roposed study[5,6]. On approval/deemed approval by the regulatory authority, the etailed study takes off with the following work elements as proposed in form 1.

- Literature survey
- Project description
- Desk research
- Baseline data generation
- Prediction & mitigation of impacts
- Post project monitoring plan
- Cost estimate

Based on the baseline data generated, the relevant environmental impacts nd proposed mitigation measures are as follows :

mpacts	Mitigation
IR	
Mine excavation Loading & unloading Wind erosion of dumps and over burden Paved and unpaved road emissions Spillage from transportation	• Water sprinkling shall be done periodically in roads within the mining area • Provision of storage of water within the mining area • Usage of high sided dumber to prevent spillage • Provision of water trough at the exists of roads for tyre washing • Green belt development around the mining area, haul roads, waste dumps & stacking space and infrastructure • Good preventive maintenance schedule for equipments & vehicles • Supply of face masks to workers and staff to prevent dust inhalation • Ambient air monitoring shall be carried out for corrective measures • Vehicles transporting the minerals shall be provided with tarpaulin cover
VATER	
Ground water depletion causing drying of wells	• Provision of storage of run-off water to facilitate ground water recharge, artificial recharge to the effected wells in the vicinity of the mining area

- Change in hydrological regime in and around the mining area
- Deterioration of surface water and ground water in and around the ML
- Mine effluent flow to the water body could affect the aquatic flora & fauna

- Design of drainage system to avoid mine flooding
- Provision of channels around the mine pit to prevent inflow of surface water during heavy rain & cyclones to prevent inundations of mine.
- Provision for settling the mine pit water and treating through neutralization tank and final discharge shall confirm to prescribed limit of inland water discharge (Figure 1).

LAND

- Sensitive location, habitations, water bodies, power lines and roads exists around proposed ML area.
- Increase in vehicular traffic
- Development of mining scar on the land surface effecting the natural beauty of the area
- Change in natural drainage pattern and the possibility of earth flow /landslide

- Plantation of the reclaimed area with species growing in the surrounding area and suiting to the ecological condition
- Green belt around the ML area, haul roads, waste dumps and infrastructure shall improve the ecological and hydrological balance, noise control aesthetics, conversion of waste land to a productive area.
- Mine pit refilling in a designed phased manner as per mine management plan approved prior to mining shall restore the topography to a better environment.
- Barren waste land could be converted to productive land after the mining activities are over.
- Disposal of overburden and waste will be done in pre-determined location with proper design and environmental planning
- Garland drains shall be provided around the boundary of dumps.
- Height alignment and slopes of the dumps shall be designed as per the wind velocity, direction and rainfall.

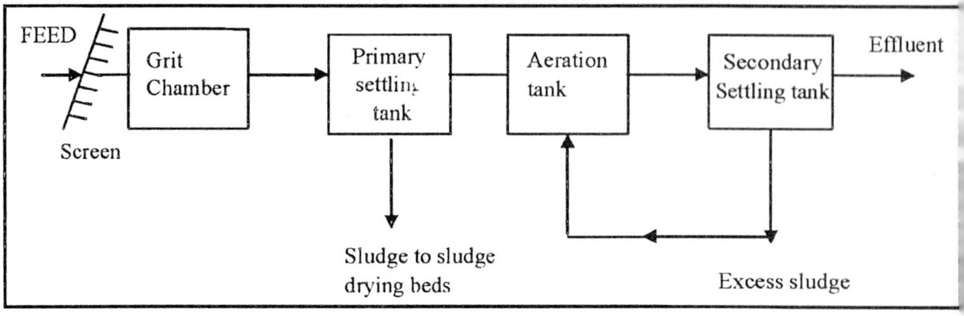

Fig. 1. Line diagram of proposed mine pit water treatment plant

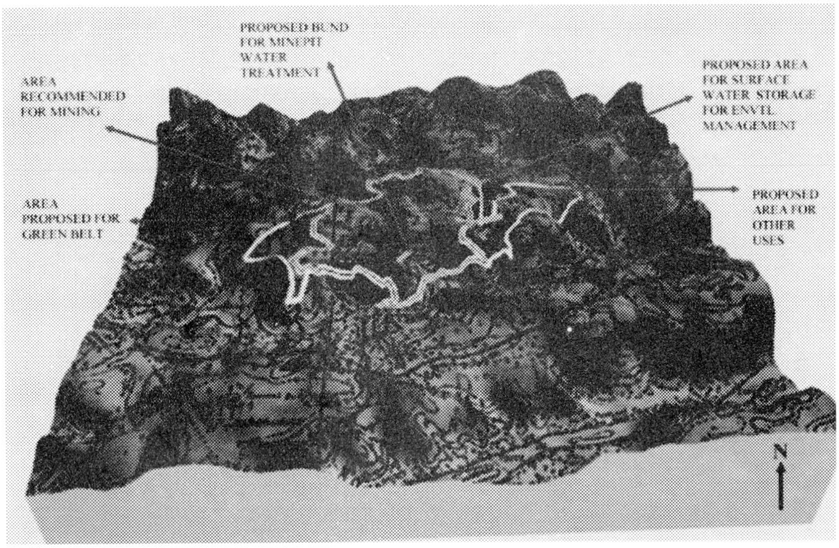

Fig. 2. Environmental management programme

Monitoring programme

The major environmental concerns for a mining project are: Air, Traffic, Water, Noise, Socio-economics and R&R and Ecology. Proper monitoring programme with schedules, equipments required for the above and manpower involvement with qualifications & cost has to be provided (Figure 2).

Acknowledgement

Ministry of Environment & Forests, Government of India, Director, NIIST and Head, Environmental Technology, NIIST are duly acknowledged.

References

1. Handbook of Environmental Procedures and Guidelines, Ministry of Environment and Forests (MOEF) Government of India.
2. Ministry of Environment & Forests notification dated 14th September, 2006, http://envfor.nic.in/
3. Preliminary environmental appraisal study for the proposed China clay project at Kasargod prepared by NIIST, Thiruvanthapuram.
4. Global Environment Outlook 2000 By United Nations Environment Programme James & James/Earthscan ISBN 1853835889, 1999.
5. EIA for Madayi Lignite and China Clay project, Kannur.
6. Environmental Impact Assessment Handbook by National Environmental protection Council.

Factors affecting the climatic changes of Assam and possible ways of recommendation

Neelima Saikia, H.P. Deka Baruah, Arup Jyoti Das, Abhijt Sarma Roy & Rosy Yein

North-East Institute of Science and Technology, Jorhat -785 006, Assam, India
[E-mail: neelimasaikia@gmail.com]

Introduction

Assam is one of the eight states of the North-East India refers to the cluster of eight states namely Assam, Meghalaya, Arunachal Pradesh, Mizoram, Nagaland, Tripura, Manipur and Sikkim. It is the eastern most region of India and lies on the lap of easternmost Himalayan hills. It has an area of about 2, 55168 km^2 with Assam occupying the largest area of 78,438 km^2. Geographically Assam is located at 24°44$'$N to 27°45$'$ N latitude and 89°41$'$E to 96°02$'$E longitude. The climate of Assam, about thirty years ago was very pleasant with an average temperature of >10°C during winter and 23-26°C during summer and was rich in biodiversity. Climatically the year is divided into four seasons, namely winter which starts from the end of October and continues till February, spring from March to middle of May, summer from May to August and autumn from September to October. The climate of Assam in general is humid in nature which is mostly attributed to the southwest monsoon which begins from the middle of June. Regarding the rainfall of NE India, Cherapunji was famous for highest rainfall in the world. The annual rainfall in Cherrapunjee during 1890s was recorded to be 475.61[1]. The average annual lowest and highest rainfall in Assam during 2006 to 2010 was recorded to be 1.2 mm to 76.93 mm. This part of the country was bountiful in natural resources with greater part of its population engaged in agrarian activities. Although Assam is very rich in natural resources, because of its geographical location and poor communication facilities from other parts of the country and world, the rapidity of industrialization in this region was slow till the Government of India adopted certain policies for the development of NE states.

Industrialization coupled with modernization in various aspects has however had a notable change in the so mentioned pleasant climate. Establishment of new industries, construction of buildings, roads and many other edifices had led to the destruction of many of the forests and valuable plants. The total area under encroachment in the forests up to 2003 has been estimated to be 3,555 km^2.

Virgin land was used up and wet lands were filled up without following any standing guidelines leading to the pollution of air, soil and water. Industries of Assam are mostly agro-based and mineral-based. The major contributors are petroleum, coal, tea, paper, limestone, cement etc. These industries directly or indirectly contribute to polluting the atmosphere for not following the guidelines for pollution control laid down for the industries. Further, the extensive use of agrochemicals for production of more food crops to meet with the demand of increasing population also generated green house gases in the atmosphere. The total population of Assam according to 2001 census is 26.66 million[2] and the average increase in population from 1971 to 1991 has been estimated to be 52.44%. This percentage clearly indicates the trend to growth and development this region has followed.

Naturally occurring resources in the region

Assam is rich in natural resources like forests, flowering plants, wild orchids, bamboo, cane, medicinal and aromatic plants, plantation crops, tea, minerals (oil and coal), water and highly fertile land.

The climatic condition in this part of the country coupled with heterogenic physiography favours the growth of a wide variety of plant species. The natural vegetation of NE India particularly Assam can be divided into evergreen forest, deciduous forest, swampy forest, grassland and riparian forest. The total forest area of Assam is around 27826 km^2 and the forest and tree cover area is around 28761 km^2.Tall trees with evergreen foliage grow in those areas where average annual rainfall is more than 200 cm and temperature is 25°C. NE India has been rich in natural vegetation with ever green forest which were full of valuable trees like hollong (*Dipterocarpus macrocarpus*), bamboo (*Bambusa* sp.), pinus (*Pinus khasiana, P. logifolia*), cryptomeria (*Cryptomeria japonica*), nahar (*Messua ferrea*), sal (*Shorea robusta*), shishu (*Dulbargia sisso),* titachampa *(Michelia champca*), bonsom (*Phoebe goalparensis*) etc.

In Assam deciduous forests were found in the districts of Dhubri, Kokrajhar, Bongaigaon, Goalpara, Barpeta, Nalbari, Kamrup, Darrang, Marigaon, Nagaon, Sonitpur, Karbi Anglong, North Cachar Hills and the drier parts of the Barak Valley where the average annual rainfall was 80 cm to 200 cm . These forests were full of plants like teak (*Tectona grandis*), gamari (*Gmelina arborea*), simul (*Bombax malabaricum*), kadam (*Anthocephalus indicus*), siris (*Pithecolobium saman*), bhomora *(Terminalia* sp.), etc. In Meghalaya these forests were found in Garo and Jaintia hills. The drier parts of Tripura, Mizoram, Manipur and Nagaland also had deciduous forest. The extent of water bodies within the forest area is 359 km^2 with 1.29% of forest cover. According to Environmental Information System-Assam, swampy/marshy areas constitute another major group of wetlands in Assam covering an area of 0.55 per cent of the total geographical area of the state.

The economy of Assam is agricultural based and the state is known for its granary of variety of medicinal and aromatic herbs and plants. About 1500 types of medicinal plants are known to exist in Assam in abundance. Among them sarpagandha (*Rauvolfia serpentine*), pippali (*Piper longam*), amlakhi (*Emblica officinalis*), hilikha (*Terinalia chebula*), bhomora (*Terminalia belerica*), arjuna (*Terminalia arjuna*) etc. are notable. Tea (*Camellia sinensis* var. *assamica*) is extensively cultivated in entire Assam. Presently Assam contributes to 51% of the total tea production of the country. Assam tea is famous for its stimulative property, flavor, bright color and taste. Cereals like paddy, wheat and some crops like potato are also cultivated over large areas. Assam Lemon is the unique product of Assam with a high percentage of Vitamin C (10 mg./100gm) besides Vitamin B (0.02 mg/100 gm), Niacin (0.1 mg/100 gm). Extractable oil content of the peel is around 0.8% and can be used in the cosmetic industry. More than 925 varieties of orchid are available in India out of which 600 variteis are grown in Assam. Aromatic Joha Rice is comparable with any popular aromatic rice in the world. Bhut jalakia (*Capsicum sinensis* Jacq.) till recently was marked as the hottest chilli of the world.

The climate of Assam is suitable for sericulture. Once upon a time traditional varieties of silk like eri, muga and pat (mulberry silk) were extensively produced in Assam. The muga silk is known for its lustre, polish and golden colour. These silk have a huge market potential locally and abroad.

The major mineral resources found in Assam are petroleum (crude oil) and natural gas, coal, limestone, iron ore, granite, fireclay, lithomarge, magnetic quartzite, kaolin, feldspar, fuller's earth, sillimanite and glass sand. Among the mineral resources crude oil occupies the top most position. It is known for its superior quality and associated natural gas production. Assam was the first state in the country where in 1889 oil was struck at Digboi in Tinsukia district. Like petroleum, natural gas is also valuable source of power and various other chemical by-products. In Assam, almost all the petroleum producing areas of the Brahmaputra Valley, especially Naharkatia, Moran, Lakuwa and Rudrasagar, contains associated natural gas. Coal is also found abundantly in the North-East, particularly in Garo hills, Khasi and Jayantia hills, Mikir hills, Margherita & Liddu of Upper Assam The coal in this region is unique, because of highly volatile (36% - 42%) quality. Ash content (3% - 15%) is low and possesses high crackling index (10% - 29%). It is friable in nature and has high sulphur content, and is utilized mainly by the railway department and power stations. Limestone occurs in Lakhimpur, North Cachar Hills, Karbi Anglong, Nagaon and Sivasagar districts. It is an important mineral used in the manufacture of cement, as flux in iron and steel production and as raw materials for chemical industries. Kaolin is found in Karbi Anglong and Lakhimpur district. China clay which is mostly found in Karbi Anglong district is a vital input for the ceramics industry.

Types of major industries causing pollution

Oil industry and pollution

In Assam, mining is at present concentrated mainly to four industrial minerals namely, coal, oil and gas and limestone. Among all the industries, petroleum industry is the biggest in Assam. It involves exploration, extraction, refining, transporting and marketing petroleum products. The largest volume products of the industry are fuel oil and gasoline (petrol). Petroleum is also the raw material for many chemical products, including pharmaceuticals, solvents, fertilizers, pesticides, and plastics. There are hundreds of active and abandoned oil fields in Assam. There are a few companies and four major petroleum refineries in operation presently.

Crude oil is a complex mixture of many different kinds of organic compounds, some of which are highly toxic to plants and animals. During the production of petroleum and other products, oxides of sulphur, nitrogen and carbon are produced. Moreover, sometimes oil industry uses sodium hydroxide during drilling for which neighbouring areas becomes alkaline in nature. Further, in oil fields, during drilling, gathering, transporting or in other operations some quantity of the crude oil is spilled out from drilling and operating sites. This spilled crude oil changes the biochemical nature of the soil. They cause serious threat to the flora and fauna of drill sites. Moreover, certain portions of crude oil gets leached out or percolate down to the rhizosphere region of the soil causing further damage to the existing vegetation. As a result no vegetation takes place in such affected sites for long time. Some photographs of air, water and soil pollution caused by oil industry are given below.

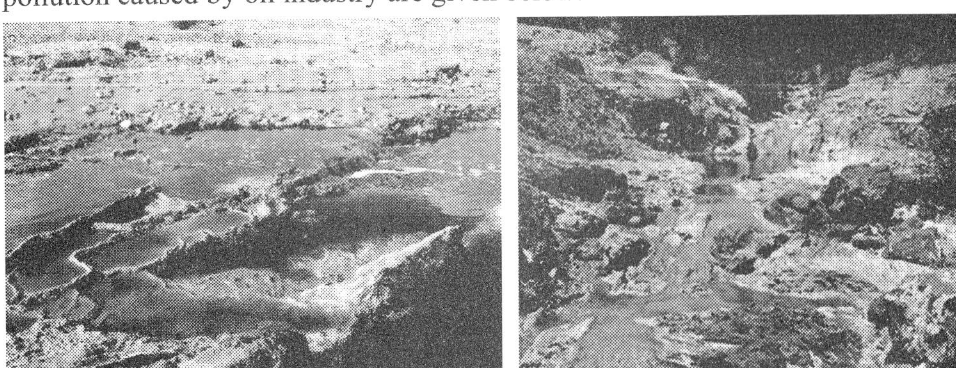

Acid drainage

Overburden soil open cast mining

Coal industry and pollution

In North-East India, Assam, Meghalaya, Arunachal, Nagaland and Sikkim are the coal producing states. Assam has large reserves of coal too. The coal of this region is unique in certain characters. They contain high content of sulphur, highly volatile and ash content is low. In Assam Makum, Margherita, Nazira, Lankadaman, Jaipur, Ledo Namdum, Koliajan, etc. areas are known for coal production. The major problem caused by the coal industry is due to open cast mining. Their preference is due to the cost effectiveness and higher yield as compared to underground mines. In open cast mining, acid drainage from mining areas to the neighboring places creates havoc. The acidic effluents discharged from the excavated pits have damaged the water sources and crops in the neighbouring areas. Also, the coal particles have left many fertile lands unproductive. The overburden which is dumped on low lying areas or hilly areas gets washed away with rain water and contributes to water pollution. There is also a high rise in cases of tuberculosis (TB), liver cirrhosis, allergy and asthma of the people working in these mines or living nearby. Some of the polluted areas due to various activities of coal industry are given below.

Tea cultivation and pollution

Tea is one of the most important and major industry which contributed to economic growth of the country. Tea industry was one of the major contributors of foreign exchange for India. But in recent years, the position of India in producing quality tea and exporting it has gone down. This can be attributed mainly due to the application of agrochemicals in the tea cultivation.

Application of synthetic fertilizer containing urea, phosphorous and potassium in large doses lowers the soil fertility and soil condition over a period of long use. Application of chemical fertilizers like urea in the soil has shown to cause serious damage in ozone layer of the atmosphere through biological process. The use of large quantities of pesticides in the tea gardens has also had a very adverse effect on the quality of soil and ground water as well. Some of the pesticides still used in this region have already been banned in developed countries long time back as they are also responsible for depletion of the ozone layer. Herbicides which are continuously used against weeds have also had ill effects on soil beneficial microflora and fauna. It also contributes to soil erosion. Besides, tea factories use furnace oil as a direct fire heater, mainly for economic reasons and this produces high amount of sulphur. Natural gas is also used in large amounts. Also various detergents and chemicals used in production machinery when released untreated contributes to water pollution.

Effect of pollution on the Brahmaputra and other rivers

The Brahmaputra is one of the largest rivers in the world containing about thirty percent of the country's total water resources and has a vast economic importance to the people of this region[3]. But in recent times the rise in human population and their consequent activities has had a devastating effect on the river. Rapid industrialization and urbanization without planning is destroying the basin areas of the river and its tributaries. Ploughing of the land for extensive cultivation and cutting of trees in the hilly regions make the land surface more susceptible to soil erosion. The tributaries of the Brahmaputra carry heavy runoff with large volume of detritus particularly form the areas where deforestation or soil erosion has taken place. It increases the riverbed level and reduces the water carrying capacity of river causing floods due to bank overflow. Also, the consequent bank spill raises the land surface in the plains. This intensifies the flood condition and leads to slit change of the river.

Pollution is one big problem which is having a devastating effect on the water of the rivers of this region. The surface runoff from urban areas carries oils, chemicals, fertilizers and many other toxic compounds and ultimately releases them into the river. This is changing the biological and the chemical properties of the river, thereby effecting the aquatic population and human beings as well. The biological oxygen demand (BOD) in the rivers has gone up considerably. The inorganic toxins enter the food chain and have already resulted in the death of many fishes and also had certain effects on human beings consuming such fishes or water of the river. One example of pollution on the Brahmaputra is the population of the river dolphin *Platanista gangetica* which have considerably gone down.

The major industries releasing untreated effluents into the rivers are the paper industries and the oil industries. Another cause of pollution of the water bodies is the drowning of idols during various festivals. Assam is basically a Hindu state and celebrates almost all the Hindu pujas involving idol worship. The releasing of idols into the rivers after worship however leads to an increase in the concentration of many elements including heavy metals as the certain paints used in the idols contains heavy metals. Also the plaster of Paris (POP) used in the idols lowers the concentration of dissolved oxygen in water.

Dwindling wetlands of Assam

Assam, once upon a time was endowed with numerous wetlands. Most of the wetlands of Assam's flood plains are of the two major rivers – the Brahmaputra and the Barak. They offer valuable economic and environmental benefits to human beings. The livelihood of many communities living nearby depended on these wetlands[4]. These wetlands sustains a variety of life forms and also supply products like fish, timber, rice, hay, fruits, seeds, berries etc for human use as well as for livestock. In terms of environment protection, they slow floodwaters, protect uplands from erosion, increase water quality and recharge groundwater. They help in the sequestration of carbon dioxide thereby regulating climate change.

Many of the wetlands of Assam at recent times have however been destroyed mainly due to human activities which involve settlement of new population, construction of unauthorized buildings/industries, improper waste-management, etc. These wetlands are subjected to harsh organic processes caused by debris from floating vegetation and silt carried by rain water. This has had a disturbing effect on the ecological balance and the natural serenity which existed earlier can hardly be seen now.

Possible ways of recommendation

In spite of the utmost care taken by the oil industry, soil water and air gets polluted due to occasional tank failure, pipeline rupture, burst or fire in the pipe line, tank and reservoir etc. To remediate crude oil contaminated sites scientists of North-East Institute of Science & Technology developed a quick bio-phyto remediation method to revive the soil condition[5]. Emissions of carbon dioxide during oil production by oil industry are quite high. Of course, petroleum industries are already considering greenhouse gas capture and storage methods to reduce their emissions to the atmosphere. Currently, CO_2 is widely used for enhanced oil recovery (EOR) projects to extract more oil out of aging reservoirs. Further, biofixation of carbon dioxide is possible by certain photosynthetic algae such as *Chlorella, Spirulina, Diatoms,* etc. As these algae are very efficient in fixing atmospheric carbon dioxide for photosynthetic activity, therefore, some technique can be used either to fix atmospheric one or trapping carbon dioxide at the source. Likewise degraded land due to opencast coal mining area, wasteland due to chemical contamination can be reclaimed by appropriate bio and phyto remediation method[6].

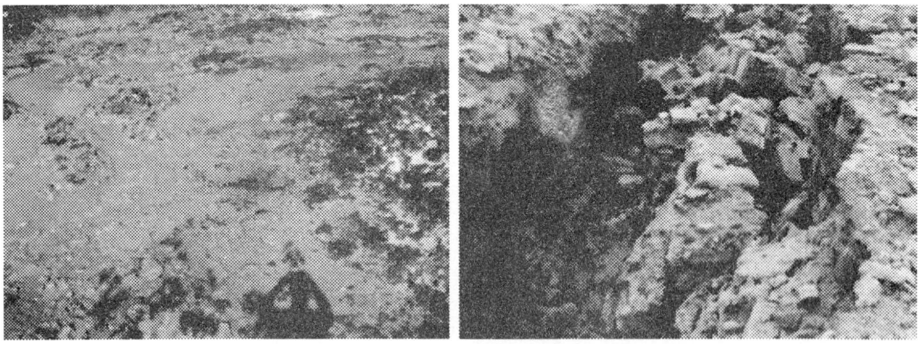

Some crude oil contaminated areas

Some contaminated areas after reclamation

It is further mentioned that to minimize and control the pollution proper monitoring should be implemented firmly to see that the industries strictly follow the environment protection guidelines. Agrochemicals affecting the concentration of green house gases (GHG) should be restricted to bare minimum and more stress should be given on use of biocontrol agents[7] and biofertlizers. Conservation of wetlands should be given priority to maintain the food chain. Big ponds should be created in each village using the traditional expertise[8] which will keep the environment cool besides being used by local people as a source of water for various purposes. To reduce the emission of GHG, use of bio-diesel, battery operated vehicle etc should be encouraged, also mass transportation system should be improved which can reduce the number of personal automobiles. After adapting all the precautions and preventive measures to control pollution if the leakage of oil or gas occurs due to some accident that can be taken care of by the environment scientists and experts of the specific field. Above all general awareness among the local masses about the benefits of environment improvement and protection will help control pollution and to reduce the effect on climate change.

Acknowledgement

Authors are thankful to the Director, NEIST, Jorhat for his consent to send this paper for publication at the International conference on climate change and for giving us the financial support to attend the same.

References

1. *Physical and Political Geography of the of Province of Assam*, A reprint from the report on the administration of the province of Assam for the year 1982-83. Assam Secretarial Printing Office Shillong (1896).
2. Singh K S (ed) People of India: Assam, *Anthropological Survey of India* Vol XV Parts I and II, Seagull Books, Calcutta (2003).
3. Sarma J N, "Fluvial Process and morphology of the Brahmaputra River in Assam, India". *Geomorphology* 70 (3-4) (2005) 226–256
4. Sharma P, "*An Overview on Wetlands in Assam*", ENVIS Assam, Assam Science Technology and Environment Council (2003)
5. Saikia N, Kumar Dileep and Deka Baruah H P, A project completion report on "*Reclamation of drill sites through application of bioformulation*" submitted to Oil Industry Development Board, New Delhi (2008)
6. Deka Baruah H P, Kumar Dileep and Saikia N, A project completion report on *Environmental cleanup and remediation of coal mining over burden sites of Tirap Collieries, Assam*" submitted to Ministry of Coal (S&T) (2007)
7. Dutta S, Mishra A K and Kumar Dileep, Induction of Systemic Resistance against Fusarial with Pigeon Pea through Interaction of Plant Growth Promoting Rhizobacteria (PGPR) and Rhizobia, *Soil Biology and Biochemistry*, vol 40, No 2 (2008) Page 452-461
8. Saikia N, Some traditional method to keep the environment cool, published in *Amar Assam*, 15th May (2004) Page 4.

Impact of climate change on water resources of Islands

P.M. Natarajan[1], N. Ramachandran[2], Shambhu Kallolikar[3] & A. Ganesh[4]

[1]Centre for Climate Change, Periyar Maniammai University, Thanjavur, India
[2]Periyar Maniammai University, Thanjavur, Chennai, India
[3]Secretary to the Governor, Rajbhavan, Chennai, India
[4]Geography Department, Bharathidasan University, Tiruchirapalli, India
[E.mail: [1]natarajan_pm@yahoo.com, [2]vc@pmu.edu;
[3]shambuias@yahoo.com; [4]aganesh2001@gmail.com]

Introduction

In many mainland countries, Island Nations and islands the snow and glacier melt is the main source of water for their different uses. In the event of snow and glacier water retreat is not arrested by controlling the carbon emission, about 1,200 cubic km of snow water in the Himalayas and about 24.5 million km^3 of fresh water in the form of snow and glacier in the land and sea area is going to be completely melted and reach the sea within 30 to 40 years from now as per the IPCC findings (IPCC, 2007 and The Water Cycle). In that case, about 2 billion people who depend on the snow and glacier water on the land area for their domestic, agriculture, industry, hydropower, fishing and recreation are going to be affected very drastically. As a result of this about Rs. 44,991 million crores ($899.82 million) worth of food grain (paddy) producing glacier water is going to be lost from the land and oceans. This water is enough to the present population of the world (7 billion) for 71,029 years, or to India (1.21 billion people) for 4.11 lakh years, or to Tamil Nadu State (72.1 million people), India for 6.89 million years, or to Chennai Metropolitan City (4.62 million people), Tamil Nadu State, India for 107.61 million years as per the World Health Organization standard (WHO) of domestic water supply (Natarajan *et al.*, 2010). Present study consists solutions to safeguard the water resources of the islands against the ongoing and the future impacts of climate change based on geological, hydrogeological, geophysical, drilling, pumping test and aquifer property study, water quality and hydrological case studies carried out in Kurusadai Island, Gulf of Mannar Marine Biosphere Reserve, Tamil Nadu State, India.

Water resources status of Islands

Hydrological and hydrogeological studies in the islands indicate that in many Island Nations and small islands there is a limited source of freshwater and some of them are experiencing water stress at the current levels of rainfall input. In Lopez and many small islands salinisation of groundwater is often a major problem, especially on low-lying islands since the extraction of groundwater is often outstripping supply (U.S. Geological Survey). Due to high tides and frequent tsunami events particularly in the islands of the Pacific Ocean there is salination of groundwater. The Atoll countries and limestone islands have no surface water or streams and are fully reliant on rainfall and groundwater harvesting. These countries are likely to face severe water scarcity due to climate change.

The well known 26.12.2004 Indian Ocean tsunami event is a standing example of groundwater salinisation by which about 15 thousand million cubic feet (TMC) or 0.42 km^3 of fresh groundwater in the South Indian and about 5 TMC or 0.14 km^3 of water in the Tamil Nadu coastal aquifers, India has been salinised. The value of this salinised water in terms of food grain production (paddy) is about Rs.750 crores and Rs. 250 crores respectively.

A vast area of the coastal regions throughout the world is going to be inundated due to the future 80m sea level rise as per United States Geological Survey as a result of glacier retreat. This scenario is likely to affect the coastal population and make them ***climate change refugees*** as well as salinise the meager quantity of the fresh groundwater resources of these islands.

Exact numbers of islands are yet to be counted. However, there are a large number of inhabited and uninhabited islands in the oceans. Hong Kong has 262 islands. There are about 20,000 to 30,000 islands in the Pacific Ocean. In the Adriatic Sea there are 1,200 islands, 16 of which are inhabited. There are 1,040 islands around Britain. In Antarctic and Australia regions there are about 8,222 islands (Wiki Answers). The groundwater resource of these islands is in danger due to the anticipated sea level rise.

Owing to the factors of limited size, geology and topography, water resources in small islands are extremely vulnerable to changes and variations in climate, especially will affect rainfall (IPCC, 2001). In most regions of small islands, projected future changes in seasonal and annual precipitation are uncertain.

Access to safe potable water varies across Island Nations. There is very good access to fresh water in countries such as Mauritius and most Caribbean islands, whereas in states such as Kiribati and Comoros it has been estimated that only 44% and 50% of the population, respectively have access to safe water. Given the major investments needed to develop storage and provide treatment and distribution of water, it is evident that climate change would further decrease the ability of many islands to meet their future requirements.

In about 9,000 inhabited islands about 700 million people, one out of ten in the world are living. In Indonesia alone 235 million and about 61.5 million in Britain are living which are the comparatively populated Island Nations (WorldIslandInfo.com).

For the food and domestic demand at the rate of 1,700 cubic metre of water per annum as per the world water resources standard, about 1,190 km^3 of fresh water (42,024.26 TMC) is needed for the people of the islands. For their domestic demand alone, as per the World Health Organization (WHO) standard, about 34.49 km^3 of water (1,218.09 TMC) is needed for the present population of these islands. For their per day domestic demand about 3.34 TMC of water (94.5 million m^3) is needed. If we assume that the people of these islands are now supplied domestic water as per the WHO standard, the domestic sewage water generation would be about 974.47 TMC or 27.59 km^3 per annum. This water has to be treated and recycled to solve the water scarcity of these islands.

It is highly doubtful whether so much water is available in the islands. The exceptional Island Nations may be Indonesia, Britain, Sri Lanka, Japan, Seychelles, Taiwan etc., have sufficient water resources. But many islands are importing food grains, vegetables, fruits, meat etc., for their food demand and also other water embedded products, and in this way these islands are able to meet their water security. Now many mainland countries where there is a limited water resource buy water hidden food and other products and save their meager water resources. Under 1990 Agreement signed on 24 November 1990 between the Public Utilities Board (PUB) of Singapore and the Johor state government of Malaysia, Singapore was allowed to construct a dam across Sungei Linggui to facilitate the extraction of 250 mgd (million gallons per day) of water from Johor River. In this way Singapore is solving its water demand (Chew, 2009). So the water scarcity islands can follow Singapore.

Case study in Kurusadai Island to suggest the approaches to improve the water resources of islands

To suggest the approaches of arresting the water supply and demand gap of the water scarcity Island Nations and islands, water resource studies have been undertaken in Kurusadai Island, Tamil Nadu, India.

From the dawn of the Marine Biological study, Kurusadai Island is considered as the ***"Biologist's Paradise"*** since it is enriched with varieties of corals, sea weeds and sea animals. There are above 20 varieties of corals, out of which 9 varieties are considered as the most important and prominent. There are three groups of sea weeds, which include about 64 genera and 104 species. Because of the rich protein value, algae like ulva is used as food and chemicals are also extracted from certain sea weeds. Since this island has rocky, sandy and muddy formations, it facilitates for the growth of three habitats of sea animals and hence there is large number of sea animals surrounding this island. The most important and interesting sea animals that are worth mentioning are: (1). Dolphins (Cetaces) (2). Green turtle (Cholonemydas) (3). Ptychodora (Balano-glossues) (4). Sea cows (Sirenia).

In view of the above biological importance every year, thousands of college students, teachers, scholars and visitors are visiting this island. Further, Kurusadai Island comes under the Gulf of Mannar Marine Biosphere Reserve.

Hence, there is a heavy demand for fresh potable water in this island. Therefore, detailed studies such as geological mapping, hydrogeological and geophysical surveys, hydrology, rainfall data collection, drilling, water level and water quality monitoring, pumping test and aquifer parameters calculation etc., have been carried out in this island, to study the hydrogeological condition of the formations and to quantify the fresh ground and surface water resources and their quality.

Location and extent

There are 21 islands in Gulf of Mannar. In Mandapam area, there are 7 islands:

1. Muyal (Hare) 2. Manoli 3. Manoliputti 4. Pullivasal 5. Poomarichan 6. Kurusadai 7. Shingle. In Keezhakkarai another 7 islands are there: 8. Anaipar 9. Vallimunai 10. Poovarasanpatti 11. Appa 12. Thalayari 13. Valai 14. Mulli. In Vembar area there are 3 islands: 15.Upputhanni 16. Puluvannichalli 17. Nallathanni and In Tuticorin area there are 4 islands: 18. Van 19. Kasuwar 20. Karaichalli 21. Vilanguchalli. In these islands there is no human inhabitation but people are going to these islands every day mainly for fishing.

These small string islands are located off the coast lying roughly parallel to the Thirunelveli and Ramanathapuram coasts, over a length of about 80 miles (Indian reef region). Among these islands Kurusadai island is an important one, and is located on the south-western side of Rameswaram, and on the southern side of Pamban Pass in the Gulf of Manner in the north Latitude 9°15'00" and east longitude 79°12'00" in the 58 0/3 and 58 0/4 India Topo sheet (Plate 1).

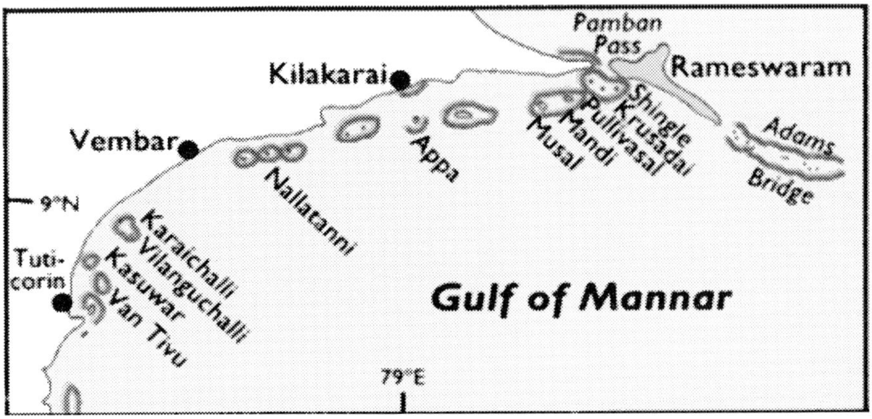

Plate 1. The approximate area of this island is 120 acres or 4, 85,640 m^2.

Location of Kurusadai and the string islands in Gulf of Manner-Rameswaram and Tuticorin coast, Tamil Nadu, India (Plate 1).

Climate, rainfall and crops

This island experiences tropical climate throughout the year and it receives major portion of rainfall during north-east monsoon period (October to December).

Pamban is the nearest rainfall station and the average 10 years (1998 –2007) annual rainfall of this station is 875.50 mm.There is no cultivation in this island. There are some trees such as Palmyra, Coconut, Pandamus, Casuarinas, Thespesia, Morcosa etc.

Physiography and drainage

Kurusadai Island is gently slopping towards the western side. There are small sand dunes which are mainly composed of powdered shells. The maximum height of this island is 2.09 m, above MSL. Rain water drains this island from east to west and stagnates for a few days and then gradually drains into the sea.

Geology

Kurusadai Island is covered by wind blown and wave borne crumbled and powdered shells of corals and mollusks with a little sand. The area covered by such powered heap of shells in this island is about 40 acres or 1,61,880 m^2.

Geophysical survey

To study the subsurface lithology and the quality of groundwater, shallow depth probes (D.P.) have been carried out in 8 locations in this island with an Alternate Current operated resistivity equipment-Terrameter. Of the 8 soundings, excepting D.P. No. 6, all are not favorable either due to low resistivity value or due to high resistivity value indicating the presence of saline water or hard and non-porous calcareous sandstone/coral limestone respectively (Table 1).

Table 1. Data showing the results of depth probes

S.No	Description	D.P. No	P1	P2	P3	P4	P5	P6	H1	H2	H3	H4	H5	H6
			Apparent Resistivity values in Ohm metre						Thickness of layers in metre					
1	Kurusadai Island	1	1050	47700	14900	-	-	-	2	40	100	-	-	-
2	-do-	2	66.5	L.I	V.H	700	L.I	-	1	10	24	28	40	100
3	-do-	3	0.42	L.I	0.25	L.I	V.H	-	5	14	16	21	100	-
4	-do-	4	6.3	126	V.H	252	L.I	157.5	1	7	12	35	60	80
5	-do-	5	0.3	1.89	5.67	1.89	-	-	2	10	22	30	-	-
6	-do-	6	2.24	44.1	182.7	-	-	-	2	9	100	-	-	-
7	-do-	7	1.57	7.87	25.2	V.H	-	-	1.5	6	14	45	-	-
8	-do-	8	3.15	18.9	V.H	L.I	V.H	-	2	7	13	33	100	-

Even in the location (D.P. No. 6) where moderate resistivity value was obtained, it was found that after constructing a shallow bore well to a depth of 37 feet, the quality of the groundwater was highly saline, indicating the non-availability of fresh groundwater.

Drilling

A shallow bore hole was drilled to a depth of 37 feet in this island and the following formations are met with:

Depth in feet	Formation
Ground surface to 10	Powered shells and corals
10 to 20	Powered coarse grained shells and corals
20 to 26	Powered coarse grained shells and corals with a little sand and clay
26 to 32	Fossiliferous sandstone with streak of coral limestone
32 to 37	Medium to coarse grained fossiliferous sand stone.

From the above surface and subsurface lithological studies it is seen that this entire island is not made up of only by massive coral limestone but mixed with shells of marine dead animals.

Hydrogeology and aquifer characters

To study the behaviour of the formation water to rainfall intensity, groundwater level measurements have been carried out in the monitoring bore well 82161 constructed in this island and a hydrograph has been drawn for this well (Plate 2). From this hydrograph it is seen that there is raise of groundwater level from the first week of November due to the precipitation during this period. The groundwater level slowly declines from the last week of November onwards when the intensity of rainfall decreases. In this above well, during the rainy days the water level reaches to the ground surface and it goes down to a maximum depth of 1.05 m below ground surface during the non rainy days.

In the above bore well, a pump test was conducted for 150 minutes to find out the aquifer characteristics. The discharge was maintained at 16 gallons per minute. The initial and the maximum water levels observed in this well during pre and at the end of pump test are 1.48 m and 2.13 m respectively. So the maximum draw down observed was 0.65 m.

The drawdown and recovery groundwater level data have been interpreted using semi log graphs and the aquifer characteristics of the formation have been calculated and given below.

Transmissibility	:	18,900 gpd / ft
Permeability	:	540 gpd / Sq.ft.
Specific capacity	:	8 gpm / ft. of draw down

The aquifer properties reveal that the formations of this island have good transmitting and conducting properties.

Water resources

The surface and groundwater resources for Kurusadai Island have been calculated by rainfall run off coefficient and water level fluctuation methods respectively.

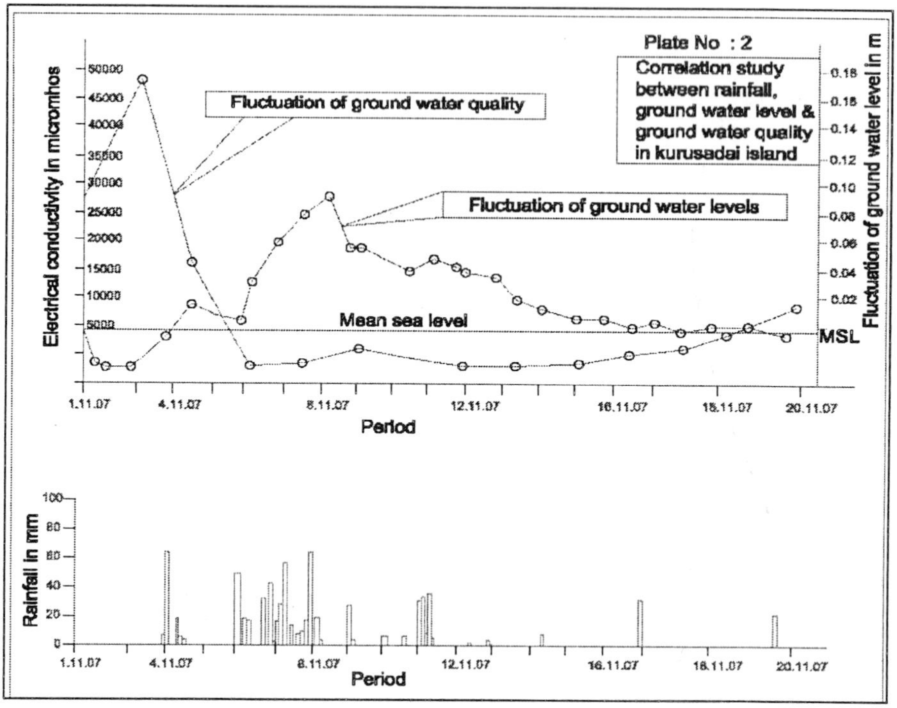

Surface water potential

As per the soil moisture accumulation studies carried out for the Pamban rainfall station for 20 years from 1980 to 2000 the average percentage of runoff, evaporation arrived by Dry-Damp-Wet method is 28.87% and 34.85% respectively. Hence, on the safer side only 15% of the rainfall is taken as the run off coefficient and the surface water resource has been calculated. The surface water potential is 62,083 m^3. Hence, this rain water can be stored in suitably designed storage tanks and utilized.

Groundwater potential

Maximum water level raise above mean sea level (MSL) in the observation well was 0.88 m. To obtain good quality groundwater and to avoid sea water intrusion, the groundwater level has to be maintained always above MSL in the coastal area aquifers according to Ghyben–Herzberg principle.

Hence, 0.2 m is left out as the dead safe water storage above MSL and only the remaining 0.68 m of the saturated zone is considered for the present extraction. By assuming 15% as the specific yield of this formation, the groundwater potential worked out by water level fluctuation method is 49,535 m^3.

Groundwater potential of sand dunes

Groundwater potential of the sand dunes in Kurusadai Island has also been calculated by water level fluctuation method. The area covered by sand dune in this island is about 40 acres or 1, 61,880 square metre. Groundwater potential for this sand dune area is 16,512 m^3.

Groundwater quality

To study the quality of groundwater, groundwater samples were collected periodically from the dug well located in Kurusadai Island and chemically analyzed. From this data it is found that the quality of the groundwater is highly saline in the dug well and also in the sand dunes (powered shell heaps) except during a short period after the rain. The electrical conductivity is very high and it exceeds above 15,000 micromhos/cm. at 25°C in many tests. The chemical type of the groundwater from these wells is Sodium Chloride (NaCl) type, indicating the mixing of the rain water with sea water since the groundwater level is below mean sea level due to the low elevation of this island. The total dissolved solids, Sodium Absorption Ratio are also high. To examine whether there is improvement in the quality of the groundwater in this Island, a correlation study has been made between rainfall, water level and groundwater quality. From this study it is seen that the quality improves only during the rainy days and becomes poor within a short period after the rain stops, as already indicated above in Plate 2.

Table 2. Groundwater sample (borewell) analysis report

		Well No: 82161
District	Rammnad	Date of collection 23.3.2007
Taluk	Rammnad	Date of receipt 6.4.2007
Village	KURUSADAI ISLAND	Date of analysis 10.4.2007

1	Temp °C (Field)	5. Specify Electrical conductance 50,800 (E.C. micromhos/cm at 25°C)
2	Odour Odorless	6. pH. Field 7.0
3	Colour Colourless	7. pH. Lab 7.1
4	Turbidity Clean	8. Total dissolved salts (TDS at 110°C) mg/L

Total dissolved salts

(a)	sum	35,100
(b)	Ec x 0.64	32,500
Total hardness as $Caco_3$	mg/L	6,800
Total Cations	Meq/L	623.5
Total Anions.	Meq/L	624.9
Cation-Anion	100	
Cation+ Anion		
Sodium adsorption ratio	58	
Geo chemical type	Nacl	

S.No	mg / L	mq / L	mg / L	meq%
1	Calcium (Ca)	260	13.0	2.1
2	Magnesium (Mg)	1496	123	19.7
3	Sodium (Na)	11040	480	77.0
4	Potassium (K)	293	7.5	1.2
5	Bicarbonate (H Co_3)	848	13.90	2.2
6	Sulphate (So_4)	1134	23.5	3.8
7	Chloride (CI)	20866	587.5	94.0

Suitability: Unsuitable for domestic, industry and irrigation

In the shallow tube well constructed in this Island also, the quality of the groundwater is highly saline and it is unsuitable for any use. The electrical conductivity value is 50,800 Micromohs /cm. at 25°C. The chemical type of this groundwater is NaCl –Table 2.

Findings from Kurusadai Island

From the above studies it is seen that, there is enough groundwater potential in the Kurusadai Island during the North–East monsoon after each rainfall. But this quantity has to be tapped, stored as quickly as possible after each precipitation since the fresh groundwater either seeps into the sea or the saline water of the sea contaminates it within a short period of each rain as the aquifer is located below MSL. Surface water potential of this island is 62,083 m^3 and the groundwater potential is 49,535 m^3. This potential is not available throughout the year as the quality deteriorates within a week after such rain and it becomes unsuitable in the remaining period. Hence, by constructing shallow dug wells in the sand dunes, groundwater can be tapped and stored in suitable structures after each rainfall for future use.

After each good rain about 62,083 m^3 of surface water drains into the sea as a waste. This water can also be stored in suitably designed subsurface storage tanks with bottom cement lining. Periodical cleaning of tanks is necessary to keep the water ever clean so that to use it continuously. If the surface water is stored in the ponds and shallow water bodies (to be constructed in future) in the low lying areas of the island it is also possible to push the saline groundwater and to improve it in stages as per Ghyben–Herzberg principles. For such islands, this is the only and most suitable artificial groundwater recharge method to improve the groundwater quality.

Conclusions

It is suggested to adopt the following strategies to protect the water resources and ecosystem of the Island Nations and islands. 1. Rainwater harvesting both from the roof and land area 2. Artificial groundwater recharges 3. Treating and recycling the domestic and industrial waste water 4. Adopting micro irrigation practices in the islands where there is agriculture 5. Desalination of saline or seawater to the domestic demand of the island people. Enhance the price of water to reduce the water use. Virtual water import to the water scarcity islands to meet food demand and other water embedded products 8. Developing groundwater in the coastal area aquifers of the islands in such a way, by keeping one foot fresh groundwater column above MSL.

In Kurusadai Island there is only floating population of about 200 people per day. Their water requirement is likely to be around 10,000 litres per day. By tapping and storing the groundwater as well as the surface water as quickly as possible after each rain, it will be possible to supply the required quantity of water for the domestic demand to the floating population. Since the drinking water demand may be around 800 to 1,000 litres per day, this small quantity can be transported from the mainland and supplied. This approach is applicable for the Island Nations and islands also which have similar hydrogeological conditions like Kurusadai Island.

Acknowledgement

Authours thank Mr. Sankar Urali, Software Engineer, USA, Mrs. Abirami Pachamuthu, Quality Assurance Engineer, Mitech, USA and Mr. Godwin, Assistant Professor for helping us to prepare the plates and tables.

References

1. Chew, Valerie 2009, National Library Board Singapore, Singapore-Malaysia water agreements.
2. Indian reef region, Gulf of Mannar and http://www.reefindia.org/gulfofman.htm.
3. IPCC Fourth Assessment Report: Climate Change 2007: Working Group II: Impacts, Adaptation and Vulnerability.
4. Is seawater intrusion affecting ground water on Lopez Island, Washington? U.S. Geological Survey, Fact Sheet 057-00.
5. Natarajan, P M *et al.*, 2010, *'Impact of climate change on the global water, food and economy'*, National conference on Biodivesity for Sustainable Development; August 25–27, 2010, organized by Centre for Biodivesity and Forest Studies, Madurai Kamaraj University, Madurai.
6. The Water Cycle: Water Storage in Ice and Snow, USGS, Science for a changing world.

Preparing for an uncertain climate — Case study over a coastal station

Bindu G.

Department of Atmospheric Sciences,
Cochin University of Science & Technology, Cochin - 682 016, India
[E-mail: mailtobindu@yahoo.com]

Introduction

It is now recognized that the global climate is warming and south Asia is especially vulnerable to impacts of changes in temperature, precipitation patterns, storm behavior and sea level rise. Coastal areas are particularly vulnerable to climate change impacts. Impact concerns for Cochin are related mainly to possible changes in precipitation pattern and a rise in sea level, with the impacts more likely to emerge gradually over the next half century, than to have a major near term effect.

Cochin is a city of about 6 lakhs people with 2 lakhs within metropolitan area. Apart from this a major contribution is from floating population. The area is warm and humid, with two monsoon periods annually, neither as intense as the major summer monsoon on India's east coast and neither associated historically disastrous flooding or other storm surge. Total precipitation is about 3000 mm annually about two third of that from south west monsoon. The average elevation of the city is about 1.5 m, crisscrossed by back waters. The climate is dominated by heavy rainfall during South west monsoon season and moderate rainfall during post monsoon season. Being a tropical coastal station, the annual range of temperature is small and there is influence of land and sea breezes. Winds are generally light to moderate throughout the year.

The effects of climate change on earth's environment will be manifested most directly through four mechanisms. First, although some climate anomalies are to be expected, most climate experts agree that human induced climate change will increase average global atmospheric and sea surface temperatures. Second, temperature increases will be associated with changes in precipitation pattern. Third, changes in storm tracks and intensities are likely to occur. Fourth, the sea level is expected to rise due to a combination of thermal expansion and the melting of mountain glaciers and polar ice cap. It is widely accepted that the climate change is likely to be accompanied by increased climate variability and extreme events. The study is an appraisal of the meteorological features of the city from different studies done for Cochin. This will provide the basic climatological information for practical applications.

Materials and methods

Surface meteorological data since 1901-2002, obtained from IMD has been used for ascertaining the meteorological background of the region. The data used is for different durations for different analysis. Different meteorological aspects like: trends in rainfall and temperature pattern, intensity of rain spells, frequency of storms etc were analyzed.

Results and discussion

Temperature

Forecasts of global climate change (IPCC 2001a) show expectations of an increase in global average surface temperature of 1.4°C-5.8°C by 2100, with median projections in the range of an increase of 3°C. Locations near equator and near oceans are expected to have temperature increase less than the global average, which suggests that effects of this global change over Cochin would be below the average.

The decadal change in temperature shows an increase in the 80s and 90s. From Figure 1, reveals that the maximum temperature anomaly is more pronounced in 80s and 90s. In 50s the dev from mean was from 0.5°C in May to 1.1°C in September. But in 80s it is 1.2°C in June to 2°C in December. In 90s it is 1°C in January to 1.8°C in September. So on an average the increase from mean maximum temperature has increased from ~1°C in 50s to ~2°C in 80s and 90s. ie; an increase of 1°C during the last 4 decades. A comparison of the averages for the periods 1983-1992 and 2000-2002 ie; the last decade also show the same trend (Table 1).

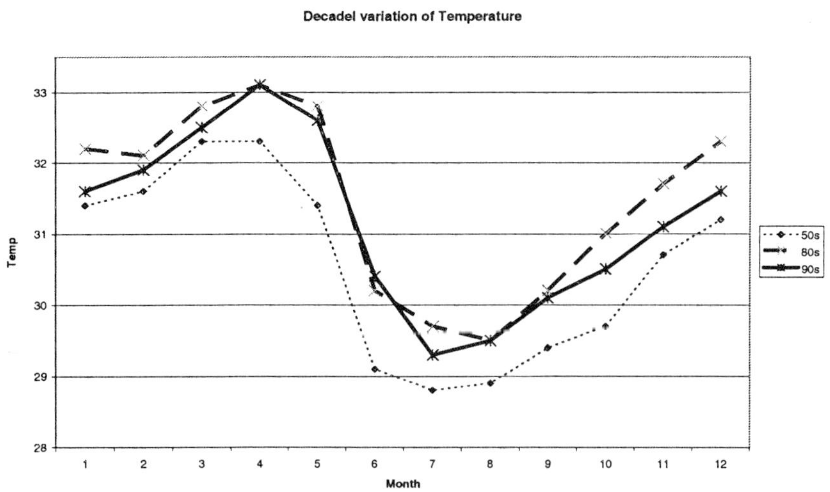

Fig. 1. Decadal variation of temperature

The three hourly air temperature values for all months based on the analysis of secondary data shows March-April the hottest months and May shows large variations in temperature values, which is attributed to the frequent pre monsoon thundershowers occurring in this month. Monthly mean values of the daily maximum, minimum and mean of temperature is given in Table 2. It shows the highest maximum temperature of 33°C occurring in April and lowest minimum temperature of 22.5°C occurring in January. Diurnal range decreases from January to June and increases from August to December. In July the month of maximum rainfall, the temperature range shows higher magnitude than that of other months. Mean temperature value during peak monsoon period is more or less the same as that of winter.

Table 1. Comparison of decadal change

Month	Mean (1983-1992)	Mean (2000-2002)
January	27.08	27.5
February	27.73	27.69
March	28.76	28.79
April	29.58	28.62
May	29.02	28.75
June	27.21	25.98
July	26.3	26.17
August	26.13	26.15
September	26.76	26.99
October	27.08	27.35
November	27.28	27.54
December	27.35	27.51

Table 2. Monthly mean maximum, minimum and mean temperature

Month	Maximum	Minimum	Range	Mean
January	31.94	22.41	9.53	27.18
February	31.96	23.87	8.09	27.92
March	32.61	25.36	7.25	28.99
April	33.09	26.01	7.08	29.55
May	32.58	26.22	6.36	29.2
June	30.45	24.63	5.82	27.54
July	29.77	23.88	5.89	26.83
August	29.35	23.91	5.44	26.63
September	30.1	24.13	5.97	27.12
October	30.87	24.23	6.64	27.55
November	31.26	23.6	7.66	27.43
December	32.31	23.39	8.92	27.85

In order to know the effect of urbanization in air temperature, heat island study was carried out from various points in and around the city center. Hourly air temperature measurements were done at different points at the outskirts of city region throughout day and night in winter months. Temperature measurements were one within the city for the same days through mobile survey starting from midnight to 8 am. The difference in temperatures between each mobile observational site and the corresponding temperature value for Cochin from IMD is noted. The study has revealed that there exists a pocket of urban heat island with intensity 2-3°C, covering most of the central regions of the city. The result when compared with an earlier study found that the heat island has now considerably spread to a larger area of the city.

Over the past 50 years, Cochin has experienced an increase in its mean annual temperature of the order of 1°C. If this rate of increase were to continue for the next 50 years, the mean annual temperature would be another 1°C warmer than present. Moreover, the proximity to sea causes city's climate to be affected by sea surface temperature. In the case of SST, if the present rate of increase continues the sea surface off Cochin coast will be warmer by 0.5°C by the end of 2050. A small increase in SST is likely to affect the local climate more than the impact of air temperature.

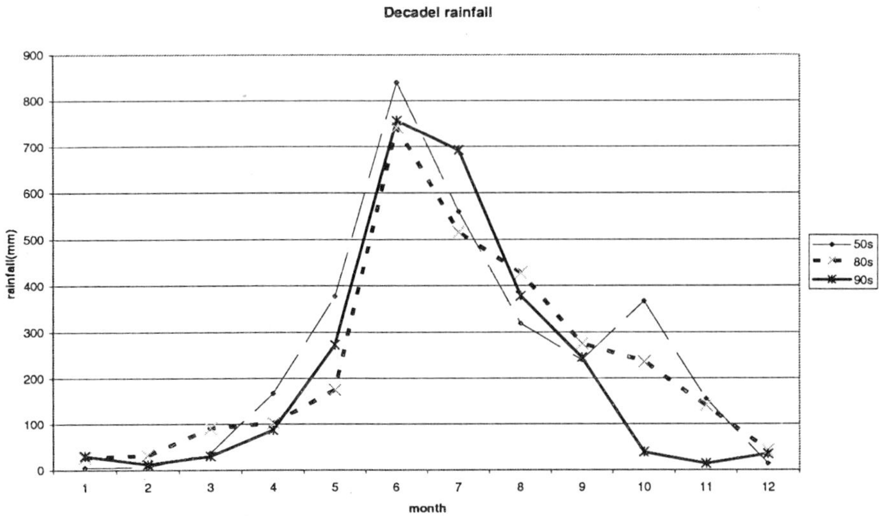

Fig. 2. Decadal variation of Rainfall

Rainfall

One of the most significant effects of global climate change is expected to be changes in precipitation pattern. Average annual rainfall recorded from 50s to 90s ranges from about 2600 mm in 90s to 3500 mm in 60s. For Cochin, rainfall pattern analysis is shown in Figure 2. Annual rainfall shows a decreasing trend for the last four decades with an exception of the month August showing a slightly increasing trend which is contributed by 80s rainfall data. If we compare the rainfall of last four decades, it shows a considerable decrease of the order of 30-50 mm. Pre monsoon and south west monsoon season also show the same tendency with the rest of the months having more or less the same trend. There is a slight increase in 90s annual rainfall which is contributed by June-July months rainfall of 90s. The rainfall was above normal during these months. The rainfall data from 1998-2002 also shows a decreasing trend, except for a slight change in the year 2001, Table 3.

Table 3. Rainfall 1998-2002

Month	1998	1999	2000	2001	2002
January	0	0	10	28	6
February	0	55	255.5	41	0
March	0	21.6	33	1.6	48
April	70.6	92.2	43.76	99.7	98
May	253.6	493.8	120.4	243	479.8
June	735.5	630.6	506.7	653.5	350.2
July	554.8	511.5	177.8	535.5	216
August	323.6	189.2	498.8	279.6	592.4
September	770.7	104.1	161.2	219.5	59.6
October	520.2	624.5	66.1	431.5	
November	78.8	38.4	39	80	58.6
December	42	0	22.2	3	27.2
Annual	3349.8	2760.9	1934.5	2615.9	1935.8
Soth west monsoon	2384.6	1435.4	1344.5	1688.1	1218.2

Cochin experiences a comparatively high intensity rainfall (Rajan, 1988). The rain spectra of station were analyzed for four years by observing the intensity per minute of the rainfall. A total of 16,000 minutes of rainfall were analyzed. Minute intensity rain spells categories are shown in Table 4. We can see that the high intensity rainfall of >120mm/hr is occurring in south west monsoon period, even though the frequency is less this has to be taken care of during city planning *e.g.* for drainage design. The rainfall that exceeds 100mm/day for the period from 1980-2000 is given in Table 5. There is not much change in the number of storm events during the period. Even though low intensity rainfall is more common during south west monsoon period, the intensities of the range > 30mm/hr and <80 mm/hr is contributing significantly into the rainfall pattern. In post monsoon also the pattern is the same with lower frequencies.

More importantly, the average annual number of rainy days in Cochin is 160 days, but over the past 50 years there has been a considerable decrease in the number of rainy days. It can be inferred that Cochin is receiving more short spells of intense rainfall. The most significant vulnerabilities of climate change are possible impacts on Cochin's water systems, especially drainage, waste disposal and water logging; or a combination of increased variability and intensity of rainfall and sea level rise and possible impacts of sea level rise on land use at the coastal margin, such as back water islands.

Table 4. Minute intensity of rainfall

Intensity mm/hr	June	July	August	September	October	November
<2	227	1044	1176	398	142	43
2-5	433	925	1072	376	92	47
6-10	505	848	1058	375	74	100
11-15	252	473	685	315	41	58
16-20	252	307	360	94	24	34
21-25	120	218	227	85	19	25
26-30	157	179	174	66	26	16
31-35	62	103	101	70	7	27
36-40	66	69	106	39	4	18
41-45	29	60	52	45	6	8
46-50	37	58	49	27	3	9
50-60	57	46	73	19	6	18
61-70	33	33	47	20	3	9
71-80	19	27	34	9	7	5
81-90	9	15	16	4		
91-100	4	2	8	4	1	
100-120	8	8	5	1		
>120	9	8	6			

Implications of temperature change

The most direct impact of increased temperature will be on the demand of water and electric power. Higher temperatures means increased rates of water evaporation and demands for potable water. They also mean higher demands for electricity for cooling, refrigeration etc; as standards of living in house holds and work places rise in coming decades. This will exacerbate the problem.

Implications of precipitation change

If rainfall in Cochin becomes more vulnerable, with more of rain falling relatively short, intense storm events, these changes could have a significant impact on city's water systems, especially drainage and waste disposal. More intense rainfall means that more of the rain runs off the surface rather than infiltering into the ground. The effect increases as more of the land surface becomes paved, which can increase the flooding and erosion. While the projections of rainfall are uncertain whether it is an increase or decrease in rainfall it is going to affect the area. If it is an increase it will be associated with more flooding and more rapid siltation of waterways. A decrease would be associated with greater likelihood of shortage of drinking water, increased water pollution due to less dilution and flushing, more extensive salt water intrusion and possibly increased costs of electricity if hydropower potentials are affected.

Table 5. Rainfall exceeding 100mm/day during the period 1980-2000

Year	Month	Date	Rainfall (mm)	Year	Month	Date	Rainfall (mm)
1981	June	4	133.6	1991	June	5	128
	June	5	106.5		June	6	116.5
	June	12	151.8		June	13	121
	June	18	106.4		October	11	104.5
	October	11	125	1992	May	19	102.5
1982	July	7	138		June	3	104.5
	August	7	103.2		June	8	186
1983	July	14	140.2		July	30	124.5
	August	2	117.8		July	7	166
	September	23	128.7		July	16	116
1985	May	24	142.4		July	28	132
	June	12	114	1993	June	6	111.6
1986	May	4	130.3		June	28	152
	June	28	120		July	16	104.6
	August	9	103.9		October	3	137
1987	Novemer	20	164.4	1994	July	17	106
1988	June	25	122.5		August	2	145
	July	22	112.5	1995	May	10	105
1989	July	27	125.5		May	11	19
	October	9	258.3		June	14	113.2
1990	May	18	112.7	1997	June	30	176.6
	June	6	103		July	24	154
	July	3	117.5	1998	July	13	106
	July	10	101		September	1	211.5
	October	27	118.2		October	12	104
1991	June	3	238	1999	June	9	120
	June	4	136	2000	August	24	112

Implications of combined temperature and precipitation changes

Combinations of changes in temperature and precipitation patterns have the potential to change local human and natural ecologies through changes in humanity and other interactions. Impacts on human health include changes in pest and pathogens to which people are exposed, changes in the levels of exposure, and associated public health requirements. Another example is possible impacts on animals and plants in both agricultural and urban areas. Less obvious but extremely important are possible impacts on cultural heritage preservation: historic structures, traditional patterns of activity and landscapes. Another area of concern is possible effects on fish populations and species of changes in sea surface temperature, chemistry and currents as a result of global climate change. If the fish and shellfish population on which fishing industry relies were to shift in location or shrink in size it would have a negative impact on an important part of the economy.

Conclusion

The major effects climate change in Cochin appear to be likely change in precipitation pattern, probably associated with an increased variability in seasonal precipitation and a very high probability of gradual sea level rise which if combined with coastal land subsidence, could be as high as one third to one half meter in 50 years. Temperatures are also likely to rise which combined with increased rainfall could increase humidity levels at some times of the year. Incidence of severe storms does not seems likely to increase, although more intense periods of rainfall could imply a greater risk of flooding.

References

1. Bindu G, *Meteorological Aspects of the Environment of Cochin*, Thesis submitted to Cochin University of Sci. & Tech. 1996.
2. Bindu G and Rajan C K, Nture of rain spectrum at a low latitude coastal station, *Pakistan Journal of Marine Sciences*, 2(2) (1993), 153-155.
3. *Environmental Status Report*, Kerala sastra sahitya parishad, Report submitted to Cochin Corporation, 2004.
4. Hingane L S, Rupa Kumar K and Ramamurthy Bh V, 'Long term trends of surface air temperature in India, *J. Climatol.*, 5(1985), 521-528.
5. Possible vulnerabilities of Cochin, India, to climate change impacts and response strategies to increase resilience, ORNL & CUSAT, Report submitted to USAID, 2003,
6. Pramanik S K and Jagannathan P, 'Climatic changes in India II: Temperature', *Indian J. Meteor. Geophys.*, 5(1954), 29-47.
7. Rajan C K, salient features of the rainspells at Cochin, *Bulletin of Department of Marine Sciences, University of Cochin,* 14(1988),131-140
8. Rupakumar K, Krishna Kumar K and Pant G B, 'Diurnal asymmetry of surface temperature trends over India', *Climatic Change*, 13(1994), 287-307.
9. Rupakumar K and Hingane L S, 'Long term variations of surface air temperature at major industrial cities of India', *Climatic Change*, 13(1988), 287-307.
10. Srivastava H N, Dewan B N, Dikshit S K, Prakash Rao G S, Singh S S and Rao K R, 'Decadal trends in climate over India', *Mausam,* 43(1992), 7-20.

Studies on the beach erosion/accretion and textural characteristics of sediments of Androth Island, Union Territory of Lakshadweep, India

[1]Sabu Joseph, [2]Vinod M.V. & [3]Prakash T.N.

[1]Dept. of Environmental Sciences, Univ. of Kerala, Kariavattom -695 581, India
[2]Dept. of Mining and Geology, Palakkad Division, Palakkad-678 104, India
[3]Centre for Earth Science Studies, Akkulam, Thiruvananthapuram- 695 031, India
[1][E-mail: jsabu2000@yahoo.co.in]

Introduction

Laccadive or Lakshadweep coral islands, which spread between N. Lat. 8° to 12°30' and E. Long. 71° to 74° in the Arabian Sea are situated at a distance of 287 to 483 km off Kochi (Fig. 1). These include Agatti, Amini, Androth, Bangaram, Bitra, Chetlat, Kadamat, Kalpeni, Kavarathi and Minicoy. The total area of the archipelago with lagoons, submerged banks and reefs is about 4200 km^2. These islands have about 20,000 km^2 of territorial waters and about 4,00,000 km^2 of exclusive economic zone. Geological study of Lakshadweep islands have been carried out by several workers (Sanil, 1986; Mallik, 1985; Mallik, 2001; Haridas *et al.*, 1989; Muralidharan and Praveen, 2001). Texural characteristics of Kavaratti island has been studied in detail (Shajan *et al.*, 2001). Similarly, the ocean waves and beach process in the Lakshadweep have also been investigated. Later, several studies (Baba *et al.*, 1992; Baba *et al.*, 1995) have been conducted to understand the wave climate and power potential of the seas surrounding Lakshadweep islands. Shoreline changes and erosion/accretion phenomena for islands were evaluated by others (Prakash and Suchindan, 1994; Prakash and Suchindan, 2001). During 1995, Prakash *et al.* (1995) studied the shoreline dynamics of selected islands of Lakshadweep archipelago in relation to wave diffraction. Beach monitoring studies for Androth island has been carried out by eminent workers (Revichandran *et al.*, 2001).

Coastal erosion has become critical for Lakshadweep Islands, which is facing acute shortage of land due to high density of population. The problems associated with the coastal erosion include loss of valuable beaches, agricultural lands, damage to the coastal vegetation and stoppage of fishing activities. An attempt has been made to study the beach erosion and accretion, and sediment characteristics of Androth Island to understand the beach dynamics and to estimate the seasonal beach volume changes, thereby, to delineate the areas vulnerable to erosion.

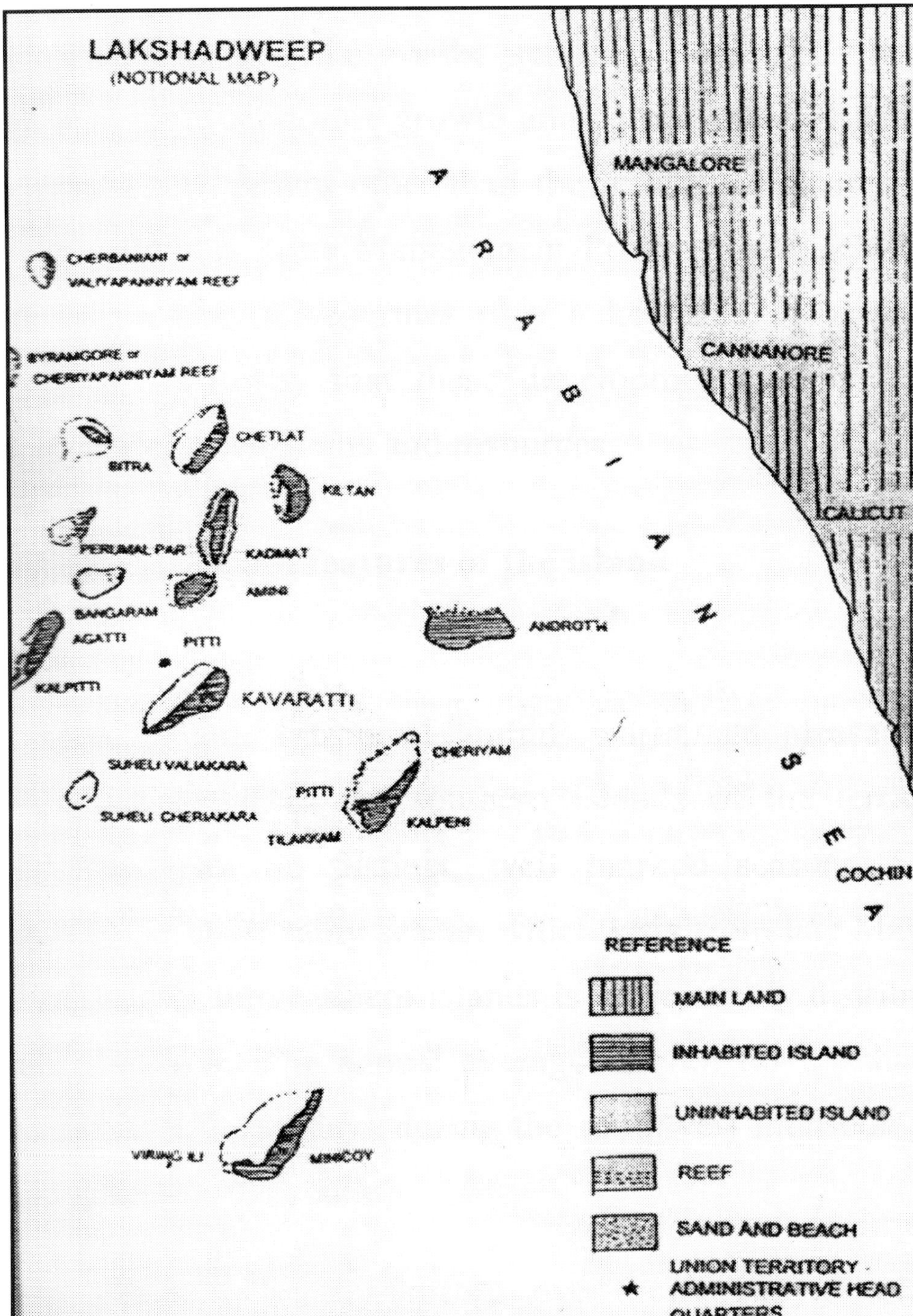

Fig. 1. Lakshwadweep Islands.

Materials and methods

Lakshadweep islands form an archipelago in the Chagos-Maldive-Laccadive ridge (Fig. 1), which is of volcanic origin (Francis and Shorr, 1966). The ridge is a major linear feature on the floor of the Indian ocean, which extend in a N-S direction for over 2500 km between 4°N and 9°N. Surface topography of the ridge is formed by shoals, banks, coral reefs and atolls, sloping downward from the sea surface to depths of 3000 to 4000 m. The ridge rises from a depth of 2000-2700 m in Lakshadweep Sea and about 4000 m in Arabian Sea. Lakshadweep islands consist of a chain of coral reefs. These reefs are mainly atoll type except one platform reef. Generally the reefs are crescent with steeper shores on the east and lagoons on the west and these reefs are of compound atoll type (Fairbridge, 1968). The islands do not show any major topographic features, but are largely low-levelled, flat-topped and generally rise to a height of hardly more than 5 m above sea level.

Most of the islands are aligned in N-S to NE-SW, except Androth which is aligned E-W. Androth is the largest one (area = 4.84 km^2; shore length = 11 km) and is roughly sole shaped with a length of 4.6 km and width of 1.5 km. It has no lagoon and occupies the whole of the atoll. It is the easternmost Islands and nearer to the west coast of Kerala. The population of Androth Island is high (=10720) and enormous developmental activities take place in the coastal area, viz., harbour development, construction of jetties etc. All these have imposed increased stress on the Island's geo-environment.

In order to study the erosion/accretion trend of Androth Island, the littoral environmental parameters (like breaker period, breaker height and direction and speed of longshore currents) and beach profile data were gathered from seven beach stations (A1-A7) around the Androth island for the months of January, August and November of 2003, and April of 2004. Among the stations, three (A1-A3) are towards the southern side and the remainder (A4-A7) towards the northern side of the Island. The profiling stations are at an interval of ~2.0 km except A5 and A6, where the interval is ~300 m due to the embayment of the shoreline. The portrait of these locations are given in Figure 2.

Breaker wave period was measured by observing the time required for ten wave crests to pass an imaginary fixed point. Time was noted by stop watch and the direction was measured using a compass. Breaker height was noted by visual observation. The heights between ten consecutive significant breaker crests and troughs are taken and the average is reported as significant breaker height. Longshore current speed and directions were measured from the reference station using neutrally buoyant floats. The alongshore distance travelled by these floats were measured using tapes. The time elapsed was noted and the speed computed. The direction was also recorded.

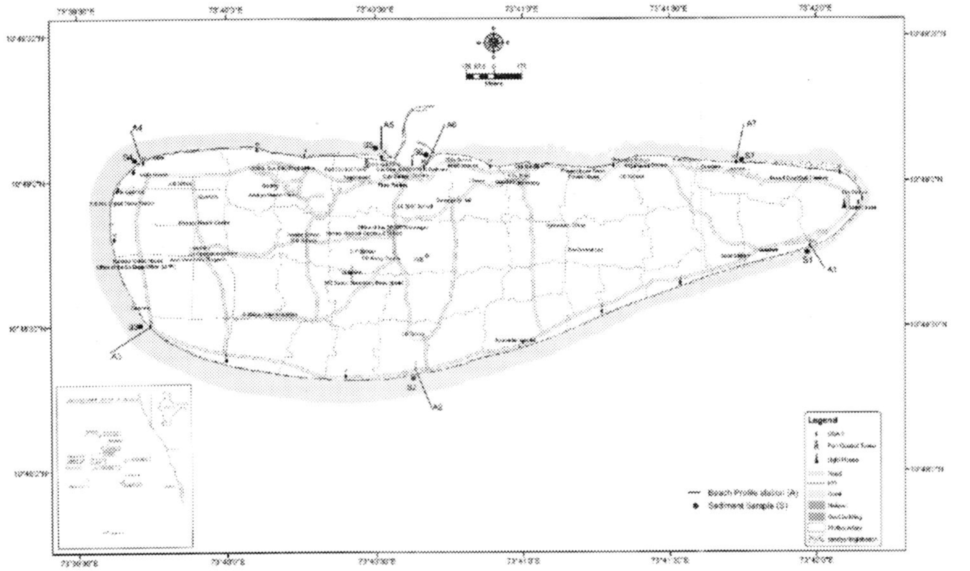

Fig. 2. Sampling stations of Androth Island

Beach profiling was done at each profiling stations around the island using a dumpy level, levelling staff and tape, and measured from the point where reference stone (i.e., benchmark) is fixed. The benchmarks are fixed based on CESS stones (CSA) and other permanent structures occurring in the backshore of the beach. After reduction with respect to the bench mark, the beach profile data were analysed by using a computer programme which compare each survey with previous survey in the same location. Thus, the program compares the profile data of each station for different periods and brings out the beach volume changes (in m^3/linear metre of the beach). The net volume changes of the beach represent the algebraic sum of the erosion in one part of the profile line and accretion in another part (Goldsmith *et al.*, 1977). Further, sediment samples (N= 7; Nos. S1-S7) were collected from each of the beach profiling stations at mid water line position (Table 1). The samples were transferred into polythene bags using a plastic spatula and were later analysed for textural attributes by sieving at half phi intervals using ASTM sieve sets in a Ro-Tap Sieve Shaker for 15 minute (Carver, 1971).

Results and discussion

Littoral processes are the result of wind, waves, currents and tides on the littoral zones. Beach morphology changes with change in the nearshore wave characteristics. Thus, a beach is constantly adjusted to wave energy and water levels. Due to inherent mobility of its sediments, it is able to maintain a dynamic environment. Waves are the principle cause of most of the shoreline changes and without wave action most of the coastal problems involving littoral processes would not occur.

Table 1. Littoral environmental parameters – Androth Island

	Jan-03				
Loc.	**BH (m)**	**Period (s)**	**Direction (°)**	**LC Speed m/min**	**LC Direction**
A1	0.70	12.5	150	1.50	E
A2	0.75	12.5	155	6.00	E
A3	1.00	12.5	235	0.60	E
A4	0.70	12	330	0.00	Oscillatory
A5	0.50	12	345	1.62	E
A7	0.50	12	345	0.00	E
Average	0.69	12.25	260.00	1.62	
	Aug-03				
Loc.	**BH (m)**	**Period (s)**	**Direction (°)**	**LC Speed m/s**	**LC Direction**
A1	2.00	9	165	17.4	E
A2	1.60	9	165	12	E
A3	1.25	9	240	7.2	NW
A4	0.75	9	335	7.8	E
A5	1.00	9	350	10.2	E
A7	1.00	9	355	7.8	E
Average	1.27	9.00	268.33	10.40	
	Nov-03				
Loc.	**BH (m)**	**Period (s)**	**Direction (°)**	**LC Speed m/min**	**LC Direction**
A1	1.00	11	160	3.6	E
A2	1.50	12	165	0	
A3	1.00	12	235	3	E
A4	0.50	11	330	1.38	E
A5	0.40	11	340	0.6	E
A7	0.50	11	350	0	W
Average	0.82	11.33	263.33	1.43	
	Apr-04				
Loc.	**BH (m)**	**Period (s)**	**Direction (°)**	**LC Speed m/min**	**LC Direction**
A1	1.00	13	155	4.8	E
A2	0.75	13	170	7.2	E
A3	1.25	12	225	1.2	E
A4	0.70	13	355	6	E
A5	0.30	12	0	0.9	E
A7	0.50	13	20	0	Oscillatory
Average	0.75	12.67	154.17	3.35	

BH= Breaker Height; LC= Longshore Currents; E= Eastward; W= West; NW= North West

The wave measurements were carried out to understand the wave characteristics of the seas surrounding the island. Breaker characteristics (Breaker height, Breaker period Breaker angle) are important in determining the quantity of sand placed in motion and the longshore transport direction and rate. Breaker characteristics measured for different periods are given in Table 1. Breaker heights (BH) show spatial-temporal variations in the study area. In general, it ranges from 0.3 (A5) to 2 m (A1). Spatial variation do exist and the stations located on the southern side of the island (i.e., A1-A3) register higher values compared to those on the northern side (Table 1). Similarly, seasonal variation also exists. Maximum values are observed in the months of August (ave.= 1.27), followed by November (= 0.82), April (0.75) and January (0.69). Breaker periods are in the range of 9 to 13 seconds (Table 2). The dominant periods are in the range of 11 to 13 seconds in January (ave.= 12.2), April (12.6) and November (11.3) respectively. Lowest period is in the month of August (9.0). However, the breaker angle (direction) does not show much variation in different beach profiling periods. A variation of only 5 to 15° is observed during different periods.

The breaking waves induce longshore currents (LC) in the littoral zone. The LC are also produced by oblique wave approach to the shore. LC is restricted mainly between zone of breaking and the shoreline, and the direction is related to the wave direction. LC velocity is determined by two factors, *viz.*, angle between wave crest and the shoreline and the breaker height. Longshore current velocity varies both across the surf zone and in the longshore direction. The transport parallel to the shore is termed as longshore transport/littoral transport or littoral drift. Longshore current speed for different periods are given in Table 1. In general, the speed ranges from 0 to 17.4 m/min. Highest values are located in A1 and A2 located towards the southern side of the island. Season-wise variations of current speed shows the highest magnitude in the month of August (ave.= 10.4), followed by April (3.3), January (1.6) and November (1.4). Androth island is oriented in the east-west direction and hence longshore currents along the southern coast were relatively stronger with velocities between 10 and 15cm/s during the observation period[14]. Direction of longshore current is mostly easterly in all the stations. No noticeable seasonal change is recorded. This is attributed to the NNW-N direction of wave breakers. Since the station A6 is entrapped between the breakwaters, littoral environmental parameters were not recorded from this station.

Beach profile is an interaction of the ground surface with vertical plane, which may extend from the top of due line to seaward limit of sand movement. Profiles are two-dimensional vertical sections showing elevations to distance variations. Coastal profiles are taken perpendicular to shoreline. Profiles measured for a given location for different periods are compared using software and based on this, net changes and cumulative volume changes are computed. Based on the cumulative volume changes, the beaches are classified into stable, eroding and accreting. Arbitrary limit is fixed to demarcate different types of beaches. If the volume changes are less than 1m^3/m of beach, the beaches are considered as *stable.* In case of *moderately eroding* or *accreting* beaches the volume changes are between 1 to 5 m^3/m of beach and when it is 5 to 10 m^3/m of beach it is considered as *highly eroding* or *accreting.* Above this limit is considered as *critical*[12]. The beach profiles of Androth island are plotted (Fig. 3) and the volume changes are calculated for each station for a period from January 2003 to April 2004, and are represented in the Table 2. The +ve and –ve values represent accretion and erosion respectively.

Table 2. Erosion/accretion (in m^3/m) trend along the Androth Island

Station	06/01/2003 to 03/08/2003		03/08/2003 to 18/11/2003		18/11/2003 to 08/04/2003		Net change (m^3/m)	Remarks
	Accretion	Erosion	Accretion	Erosion	Accretion	Erosion		
A1		-6.1		-26.36	15.74		-16.72	Critical
A2	3.9			-28.67	0.85		-23.93	Critical
A3		-2.69		-32.86		-9.25	-44.8	Critical
A4	0.83		0.65		1.42		2.9	Moderately accreting
A5	9.11			-1.28		-7.75	0.09	Stable
A6		-2.15	1.7			-10.63	-11.07	Highly eroding
A7		-2.7		-6.79	21.09		11.6	Highly accreting
Total Ero/ Acc	13.84	-13.64	2.35	-95.96	39.1	-27.63	-81.93	

The southern side of the island is represented by stations A1, A2 and A3. At station A1, the period from January '03 to August '03 shows erosion and the volume change is –6.1 m^3/m. Again, from August '03 to November '03 exhibits severe erosion and the volume change is –26.36 m^3/m. Further, November '03 to April '04 is an accreting period and the volume change is 15.74 m^3/m. The net volume change recorded is –16.72 m^3/m. Since erosion is higher than the limit, it can be grouped in the category *critical.* At station A2, the volume change for the period of January '03 to August '03, August '03 to November '03 and November '03 to April '04 are 3.9 m^3/m, –28.67 m^3/m, –44.8 and 0.85 m^3/m respectively. Net volume change is –23.93 m^3/m and can be grouped in the category *critical.* At station A3, all the periods shows an erosion trend and the station is remarked as *critical.* i.e., From January '03 to August '03 shows –2.69 m^3/m, August '03 to November '03 represents –32.86 m^3/m and from November '03 to April '04, the recorded erosion is –9.25 m^3/m. The net erosion is –44.8 m^3/m.

Beach Profile Station - A1

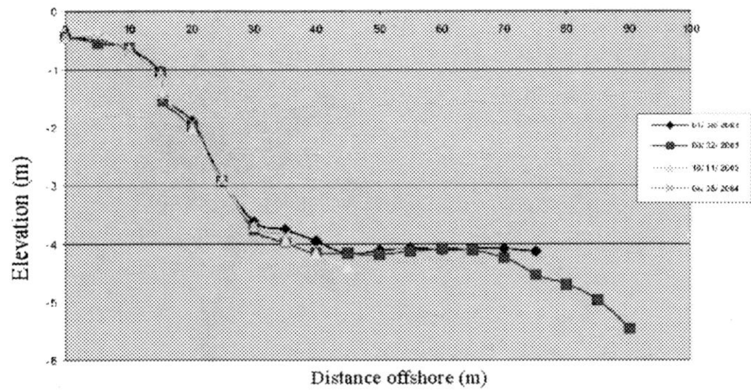

Beach Profile Station - A2

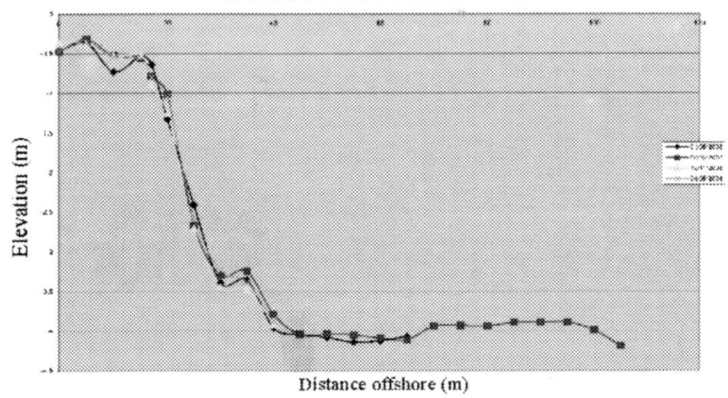

Beach Profile Station - A3

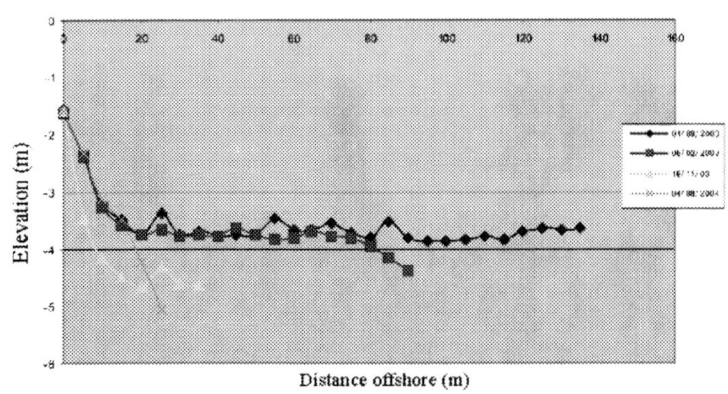

Beach Profile Station - A4

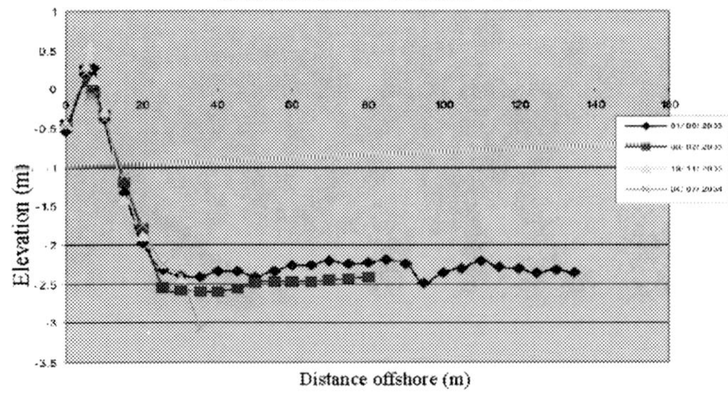

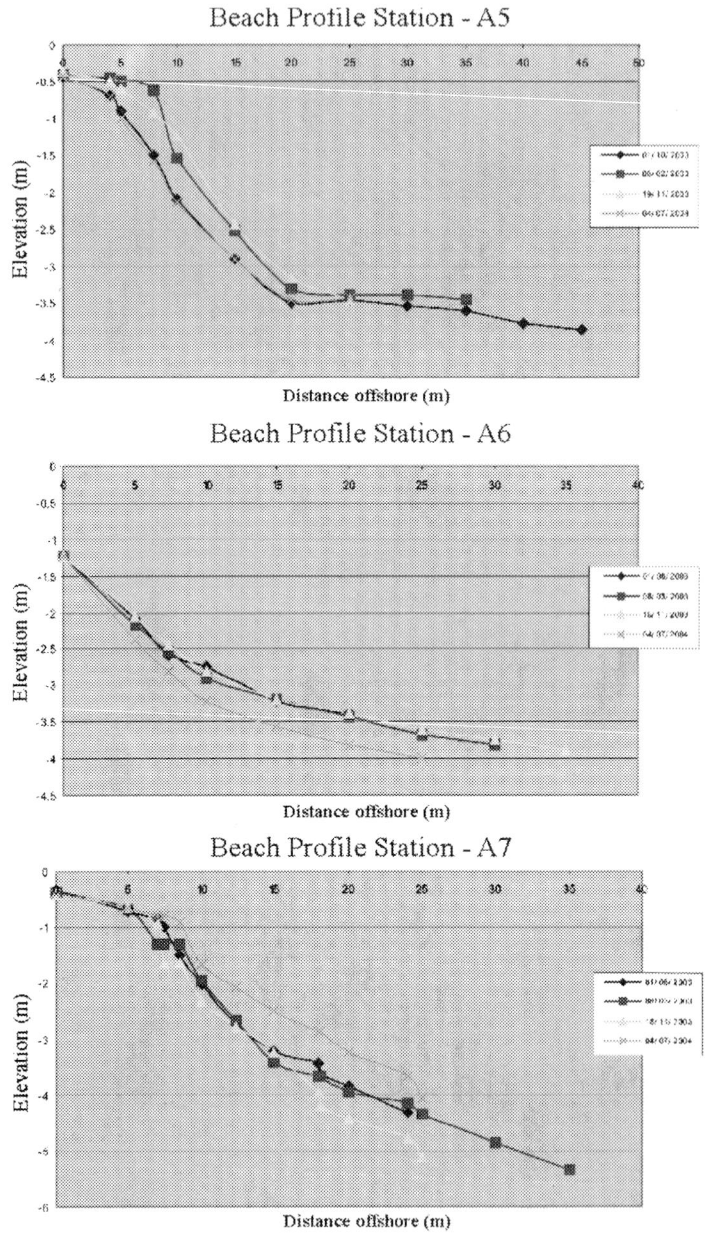

Fig. 3. Beach Profiles of stations A1 to A7

The northern side of the island is represented by A4 to A7. Beach near A4 in all the profiling periods shows accreting trend. Volume change for the period from January '03 to August '03, August '03 to November '03 and November '03 to April '04 are 0.83 m^3/m, 0.65 m^3/m, 1.42 m^3/m respectively. The net volume change is 2.9 m^3/m and is a *moderately accreting*. At A5 during the period from January '03 to August '03, an volume of 9.11 m^3/m accretion is recorded. While August '03 to November '03 and November '03 to April '04 erosion observed is

–1.28 m^3/m and, –7.75 m^3/m respectively. Net volume change is 0.09 m^3/m and the beach is remarked as *stable*. On the other hand, in A6, from January '03 to August '03 shows an erosion trend with a volume change of –2.15 m^3/m sediments. Again, from August '03 to November '03 shows an accretion pattern 1.7 m^3/m and from November '03 to April '04 shows an erosion pattern with 10.63 m^3/m. The net volume change is –11.07 m^3/m and the beach can be remarked as *highly eroding*. During the period January '03 to August '03 and August '03 to November '03, the erosion is –2.7 m^3/m and –6.79 m^3/m respectively. November '03 to April '04 is exhibited an accretion of 21.09 m^3/m. Net volume change is 11.6 m^3/m and the beach is remarked as *highly accreting*.

The stations on the southern side of the island, in general, shows an eroding trend. Out of the three stations on the southern side, the A3 recorded the maximum erosion followed by A2 and A1 with a corresponding net volume of eroded sediment –44.8 m^3/m , –23.93m^3/m and –16.72 m^3/m respectively. CSA 16 of northern side of the island recorded an erosion of –11.07 $m^3//m$. Contrary to this, the stations on the northern side of the island shows an accretion trend. At station A7 the accretion is recorded maximum at a net volume change of 11.6 m^3/m.

The southern side of the island represented by stations A1, A2 and A3 shows an erosional trend, and all of them fall under 'critical' category and warrants explanation. This is attributed to high wave activity on southern side as evidenced by relatively steeper slope, higher breaker heights, wave period and speed of the longshore currents (Table 1). In contrast, the northern side of the island is gently sloping, with a relatively lower breaker heights, period and longshore currents. Further, the break water on the northern side protects it from erosion.

The temporal variation of beach dynamics shows that wave activity is increased during SW monsoon and littoral wave parameters are high during August, and waves severely hits the southern side of the island. During August, breaker height (2m) and short period (9s) are recorded highest and is a major factor for the erosion on the southern side. High wave energy (H/T >0.25), short period of waves with near normal approach are considered to be responsible for erosion of the beach (Bhat *et al.*, 2003). The studies by CESS (1990-1993; 1997-2001) to determine the erosion of Lakshadweep islands has shown that the waves are thinning the shoreline and making the Lakshadweep islands closer to sea-level (www.cessindia.org/coastalerosion.htm). The longshore current direction in all the seasons is in easterly direction, which is attributed to the wave direction.

The changes in sea level due to global warming is another factor adding up to beach erosion and the sea level change is not globally uniform (Weiquing and Others, 2010). Studies have shown that the sea-level in the West Coast of India is rising at the rate of 1 mm per year and East Coast by 0.8 mm per year (Shetye *et al.*, 1990). In the Indian coastal region, the region most vulnerable to inundation is Lakshadeep pelago and a 1 m rise in sea level will severely restrict the human activity in this island (Shetye *et al.*, 1990). If the reefs continue to degrade by climate change and human influence, the coastal erosion would worsen (Report of the National Task Force for special study of Lakshadeep

Islands, 2005). Various management strategies are to be charted out and to reduce coastal erosion particularly on the southern side of the island.

Texture, an important grain size attribute, provides useful information on mode and extent of transport and environment of deposition of clastics (Manson and Folk, 1958; Friedman, 1961). Sediment distribution in the Lakshadweep islands is related to the source, morphology and energy of the transporting medium. The source of the sediments on the atolls is the coral reefs and other biological communities of the reef areas. In the coral island, mostly through breaking of coral debris derives the sediments. Sediments are seen in the littoral zone on which all the hydro-dynamical processes operate. Grain size characteristics of the littoral material are important parameters to be considered for any coastal engineering design and development.

Table 3. Textural attributes of sediments of Androth Island

a. Mean size

SAMPLE No.	Jan-03		Aug-03		Nov-03	
	Phi	mm	Phi	mm	Phi	Mm
S1	1.94	0.26	2.12	0.23	1.90	0.27
S2	2.01	0.25	3.25	0.11		
S3	2.03	0.24	1.87	0.27		
S4	1.71	0.31	2.26	0.21	2.51	0.18
S5	2.29	0.20	2.14	0.23	2.63	0.16
S6	2.45	0.18	3.65	0.08		
S7	1.39	0.38	2.57	0.17	3.02	0.12

b. Standard Deviation

SAMPLE No.	Jan-03		Aug-03		Nov-03	
	Phi	mm	Phi	mm	Phi	Mm
S1	0.64	0.64	0.70	0.62	0.75	0.59
S2	0.84	0.56	0.80	0.57		
S3	0.78	0.58	0.78	0.58		
S4	0.68	0.62	0.65	0.64	0.56	0.68
S5	0.53	0.69	0.53	0.69	0.57	0.67
S6	0.89	0.54	0.43	0.74		
S7	0.55	0.68	0.58	0.67	0.47	0.72

c. Skewness

SAMPLE No	Jan-03	Aug-03	Nov-03
S1	0.14	-0.12	-0.31
S2	-0.20	-0.19	
S3	-0.06	-0.12	
S4	-0.13	-0.09	-0.15
S5	0.10	0.19	0.00
S6	-0.30	0.04	
S7	0.19	-0.24	0.09

d. Kurtosis

SAMPLE No.	Jan-03	Aug-03	Nov-03
S1	1.16	0.96	0.79
S2	1.03	1.27	
S3	1.04	0.99	
S4	0.92	1.03	1.05
S5	1.36	0.85	0.98
S6	0.96	0.89	
S7	1.19	1.30	1.02

Textural parameters such as mean size, sorting, skewness and kurtosis were calculated for the beach sediments (n=7) for different seasons and are given in Table 3. Mean size of sediments, in general, ranges from 1.39 (i.e., medium) to 3.65 phi (i.e., fine). The size distribution of sediments on the northern and southern part of the island remains more or less the same and season-wise also not much variation is noticed. Standard deviation value (sorting) of sediments shows that the sediments, in general, are well (0.43 phi) to moderately sorted (0.89 phi), and there is not much spatial and temporal variation for the sorting value. Skewness ranges between –0.30 (i.e., coarse) and 0.19 (i.e., fine skewed). Kurtosis is considered as a sensitive and valuable measure of the normality of distribution. Kurtosis ranges from 0.89 to 1.36. More or less uniformity of textural parameters of beach sediments around the island indicates the common source of sediments and the minimum distance of transport of sediments. Role of energy of transporting medium in the distribution of sediments is not reflected in the textural data.

Conclusion

The study on erosion/accretion of sediments of Androth island reveals that the northern part of the island undergoes accretion, while the southern side experiences erosion. Net change observed is erosion and the volume of sediment eroded during the study period (Jan. '03 to April '04) is –81.93 m^3/m. Various management strategies are to be charted out to reduce coastal erosion particularly on the southern side of the island. More or less uniformity of textural parameters for the beach sediments around the island indicates the common source and the minimum distance of transportation of sediments.

Acknowledgments

Present study forms part of the unpublished M.Phil dissertation work done by the second author (VMV). The second author expresses his sincere thanks to Dr. N.P. Kurian, Dr. K.V. Thomas and Dr. T.S. Shahul Hameed (Scientists, CESS, Trivandrum) for the help and encouragement during the course of dissertation work.

References

1. Baba M, Shahul Hameed T S and Kurian N P, Wave climate and power potential of the seas surrounding Lakshadweep islands. Abs. V.1 Workshop. Status of Sci. Database on Lakshadweep, Geol. Surv. Ind. (1995).
2. Baba M, Shahul Hameed T S, Kurian N P and Subashchandran K S, Wave power of Lakshadweep islands. Final report submitted to DOD, Govt. of India. CESS (1992), p163.
3. Bhat M S, Chavadi V C and Hedge V S, Morphology and sediment movement in a monsoon influenced open beach at gangavali, near Gokarn. *India Jour. Mar. Sci.* (2003), 32, 31-36.
4. Carver R E, Proceedures in Sedimentary Petrology. Wiley Interscience, New York (1971), p653.
5. Fairbridge The Encyclopedia of Geomorphology- Encyclopedia of Earth Science series. V.3. Dowden, Hutchinson & Ross, New York (1968), 36, 191.
6. Francis T J G and Shorr G G, Seismic refraction measures in the NW Indian Ocean. *J. Geophys. Res.,* (1966). 71, 427-449.
7. Friedman G M, Distinction between dune, beach and river sands from their textural characteristics. *Jour. Sed. Petrol.* (1961), 31, 514-529.
8. Goldsmith V, Strun S C and Thomas V R, Beach erosion and accretion at Virginia beach and vicinity. U.S. Army Corporation of Engineers, Miscellaneous Report (1977), 77, 12, p189.
9. Haridas M G, Rastogi G K and Gupta J K, Textural studies in carbonate sands of beach and lagoonal environments, Agatti atoll. Lakshadweep. *Geol. Sur. Ind. Spl. Pub.* (1989), 24, 21-34.
10. Komar P D, Beach process and sedimentation. Prentice-Hall, Inc., Englewoo Cliffs, New Jersey (1976) p.429.
11. Mallik T K, Geology of Kavarathy and Kalpani atolls, Proc. 5th Int. Coral Reef Congress, Tahiti, (1985), 6, 419-426.
12. Mallik T K, Some geological aspects of the Lakshadweep atolls, Arabian Sea. *Geo. Sur. India Spl. Pub.* No. 56, (2001) 1-3.
13. Manson C C and Folk R L, Differentiation of beach, dune and Aeolian environments by size analysis, Mustang island, Texas. *Jour. Sed. Petro.* (1958), 28, 211-226.
14. Muralidharan M P and Praveen Kumar P, Geology and Geomorphology of Kavaratti island, Union Territory of Lakshadweep. *Geol. Sur. Ind. Spl. Pub.* (2001), 56, 9-13.
15. Prakash T N and Suchindan G K, Coastal erosion studies in Lakshadweep islands, *Proc. of Ind. Natl. Conf.* on Harbour and Ocean Engineering, Pune (1994), 31-41.
16. Prakash T N and Suchindan G K, Studies on coastal erosion and sediment movement in selected islands of UT of Lakshadweep. Final report submitted to DST, CESS (2001), p60.
17. Prakash T N, Shahul Hameed T S and Suchandan G K, Shoreline dynamics of selected islands of Lakshadweep Archipelago in relation to wave diffraction. Abs. V.1 Workshop. Status of Sci. Database on Lakshadweep, *Geol. Surv. Ind.* (1995).

18. Revichandran D, Vijayan P R, Sajeev R and Sankaranarayanan V N, Monitoring beach stability and littoral processes at Androth and Kalpeni island, Lakshadweep. *Geol. Surv. of Ind. Spl. Pub.* (2001), 56, 221-227.
19. Sanil Kumar K V, Shyam K R, Udaya Varma P and Sivada P, Topography and sedimentary characteristics of the sandy beach of Kavarathy (Lakshadweep, *Ind. Jour. Mar. Sci.,* (1986) 15, 75-77.
20. Shajan K P, Abdulla Bava K, Nambiar C G and Serlathan P, Textural characteristics of the coral sands of the Kavarathi island, Lakshadweep. *Geol. Surv. Ind. Spl. Pub.* (2001), 56, 53-68.
21. Shetye S R, Gouvia A D and Pathak M C, Vulnerability of Indian coastal region to damage from sea level rise. *Current Science* (1990), 59, 3, 152-156.
22. Weiquing Han and Others, Pattern of Indian Ocean Sea-level change in a warming climate, *Nature Geoscience* (2010), 3, 546-550.

Inter-annual and inter-decadal variability of low pressure system formed over Indian land mass and adjoining seas

C.A. Babu[1] *& Hamza Varikoden*[2]

[1]Department of Atmospheric Sciences, CUSAT, Cochin – 682 016, India
[2]Indian Institute of Tropical Meteorology, NCL, P.O., Pune – 411 008, India
[E-mail: babuca@cusat.ac.in]

Introduction

Formation of low pressure system in the Arabian Sea and Bay of Bengal brings copious rainfall though the system creates havoc to the region close to the passage. Indian summer monsoon is active during the presence of monsoon depression and both east and west coast of India get heavy rainfall due to strong moisture pumping. Since land gets continuous rainfall during active monsoon situation, the low pressure system continues to survive even after landfall during monsoon season. The variability in the number of low pressure system formed over the Indian subcontinent is important for all walks of life since variability on the rainfall pattern affect Indian economy as it is closely associated with agricultural production that depends mainly on rain fed irrigation. One of the most significant consequences of global warming due to increase in greenhouse gases would be an increase in magnitude and frequency of extreme precipitation events.

Goswami *et al.* from daily rainfall analysis found that significant rising trend in the frequency and the magnitude of extreme rain events and a significant decreasing trend in the frequency of moderate rain events over central India during the monsoon seasons from 1951 to 2000[1]. The seasonal mean rainfall does not show any trend, because the contribution from increasing heavy events is offset by decreasing moderate events. Further, Joshi and Rajeevan from an analysis for 100 stations during 1901-2000 brought out significant positive trend on extreme rainfall indices over the west coast and north western parts of Peninsular India[14]. Feng and Hu studied effect of the North Atlantic SST variation on the Indian summer monsoon rainfall at multi decadal and longer timescale and found that the North Atlantic SST anomaly strongly affects the Tibetan Plateau surface temperature and heat sources[2]. The latter causes change in the meridional temperature gradient between Tibetan Plateau and the tropical Indian Ocean. Through the thermal wind effect, the meridional temperature gradient anomalies change the lower tropospheric zonal winds. This affects the monsoon circulation and rainfall.

A study on severe droughts in India by Kumar *et al.* reported that severe droughts are accompanied by El Nino events[3]. Yet all El Nino events have not always produced severe droughts. They showed that El Nino events with the warmest sea surface temperature (SST) anomalies in the central equatorial Pacific are more effective in focusing drought-producing subsidence over India than events with the warmest SSTs in the eastern equatorial Pacific. The physical basis for such different impacts is established using atmospheric general circulation model experiments forced with idealized tropical Pacific warming.

Landsea *et al.* reported that the number of intense cyclone over the Atlantic is on the decrease[11] while Johnny and Shi found an increasing trend for the inter annual variability in tropical cyclone over northern west Pacific except during 1959-94[13]. Camargo *et al.* made an analysis on the influence of natural climatic variability on tropical cyclones and seasonal forecasts of tropical cyclone activity[8]. They reported that the response of tropical cyclone activity to El Niño-Southern oscillation and the Madden Julian Oscillation in various regions of the world depends on the genesis location, track, type and intensity of tropical cyclones. Webster *et al.* examined the number of tropical cyclones, cyclone days and cyclone intensity during the past 35 years, in an environment of increasing sea surface temperature. A large increase was seen in the number[12]. From the analyses of tropical cyclone records from the western North Pacific, Chan argued that the recent increase in occurrence of intense typhoons[10] reported by Webster *et al.*[12] is not a trend. Rather, it is likely a part of the large inter decadal variations in the number of intense typhoons related to similar temporal fluctuations in the atmospheric environment. Chan made an analysis on decadal variations of intense typhoon occurrence in the western North Pacific[9]. He concluded that the frequency of intense typhoon occurrence undergoes a strong multi-decadal (16–32 years) variation due to similar variations in the planetary scale oceanographic and atmospheric conditions that govern the formation, intensification and movement of tropical cyclone.

Hu and Feng examined multidecadal variations of the atmospheric circulation in the North American monsoon region[6]. They found that during warm and cold phases of the North Atlantic Multidecadal Oscillations, distinctive circulation anomalies are found in central and western North America, where lower than average pressure prevailed in the warm phase and the opposite anomaly in the cold phase. Associated wind anomalies configured different patterns in moisture transport and might have contributed to the development and variation of the monsoon regimes. Krishnan and Sugi analyzed Pacific decadal oscillation and variability of the Indian summer monsoon rainfall[7]. They found that the Indian monsoon is more vulnerable to drought situations, when El Nino events occur during warm phases of the Pacific inter decadal variability. Conversely, wet monsoons are more likely to prevail, when La Nina events coincide during cold phases of the Pacific inter decadal variability.

Kumar *et al.* found an inverse relationship between the El Nino-Southern Oscillation (ENSO) and the Indian summer monsoon (weak monsoon arising from warm ENSO event) has broken down in recent decades[4]. Two possible reasons emerge from the analyses. A southeastward shift in the Walker circulation anomalies associated with ENSO events may lead to a reduced subsidence over the Indian region, thus favoring normal monsoon conditions. Additionally, increased surface temperatures over Eurasia in winter and spring, which are part of the middle latitude continental warming trend, may favour the enhanced land-ocean thermal gradient conducive to a strong monsoon. These observations raise the possibility that the Eurasian warming in recent decades helps to sustain the monsoon rainfall at a normal level despite strong ENSO events. Patwardhan and Bhalme studied the trend on cyclonic disturbance formed over India and the adjacent seas and brought out a significant decreasing trend on an annual as well as in the monsoon season. They reported same trend on the decadal time scale[5].

In view of the importance of climate change scenario, research work on the variability of low pressure systems is need of the hour. Here, we made an attempt to understand the various periodicities embedded in the time series of number of low pressure system formed from 1891 to 2007 (117 years). Further, we examined nature of trend in the time series over the Arabian Sea, Bay of Bengal and for the entire region together. In addition, linear correlation coefficient analysis is made between all India rainfall and number of low pressure systems for understanding multi decadal relationship employing sliding correlation of 21 year window.

Materials and methods

We obtained the data on low pressure system (depression and cyclone) formed over the Arabian Sea, Bay of Bengal and Indian land mass from India Meteorological Department. The number of low pressure system formed over the region during each year for 117 years (1891 to 2007) is considered for the analysis. All India Rainfall was procured from Indian Institute of Tropical Meteorology for the same period and was analyzed to relate it with the number of system over the Indian land mass and adjoining seas.

We utilized wavelet decomposition method for identifying the different harmonics in the time series data. This method of decomposition can be employed where a predetermined scaling may not be appropriate because of a wide range of dominant frequencies. Wavelet analysis attempts to solve by decomposing the time series signals into time/frequency space domains simultaneously. This method of signal decomposition provides information on both the amplitude of any periodic signals within the series and how this amplitude varies with time. In this study, *dmey* wavelet technique is utilized to obtain the predominant harmonics embedded in the time series of number of low pressure systems.

Trend analysis was carried out on the time series of number of low pressure system formed during southwest monsoon season separately for Arabian Sea and Bay of Bengal. In addition, similar analysis was carried out for the total low pressure system for Arabian Sea, Bay of Bengal and land together. To examine the multi decadal relationship between number of low pressure system formed over Indian region and all India rainfall, we employed the sliding correlation for 21 year window.

Table 1. Seasonal and annual features of low pressure system formed over Arabian Sea, Bay of Bengal and Indian land mass

	Jun-Sep	Oct-Nov	Dec-Feb	Mar-May	annual
Mean	3.2	3.5	0.9	1.1	12.6
Standard Deviation	1.6	1.5	0.9	0.9	3.7
Minimum No.	0 (1913,2002)	0 (1895)	0	0	5 (1993)
Maximum No.	7 (1961)	7 (1961, 75,77,82)	3 (1923, 63, 72, 73, 2005)	4 (1927)	20 (1925, 75, 82)

Results and discussions

In order to understand the variability in the number of low pressure system formed in different seasons as well as annual, we present the statistics for 117 years on the Table 1. The average number of low pressure system formed in a year is 12.6 with a standard deviation of 3.7. Maximum and minimum number of system were 20 (in the years 1925, 1975 and 1982) and 5 (in the year 1993) respectively. Highest frequency of occurrence of low pressure system is during post monsoon season (on an average 3.5 system per year) followed by southwest monsoon season (3.2 system per year). Lowest frequency is during winter (0.9 system per year) followed by pre-monsoon season (1.1 system per year).

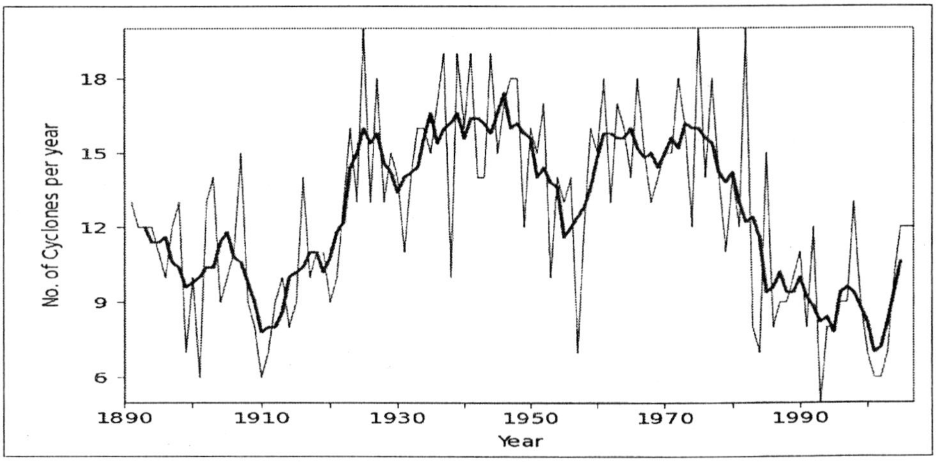

Fig. 1. Time series of total number of low pressure system on an annual basis. The bold curve represents 5-year moving average

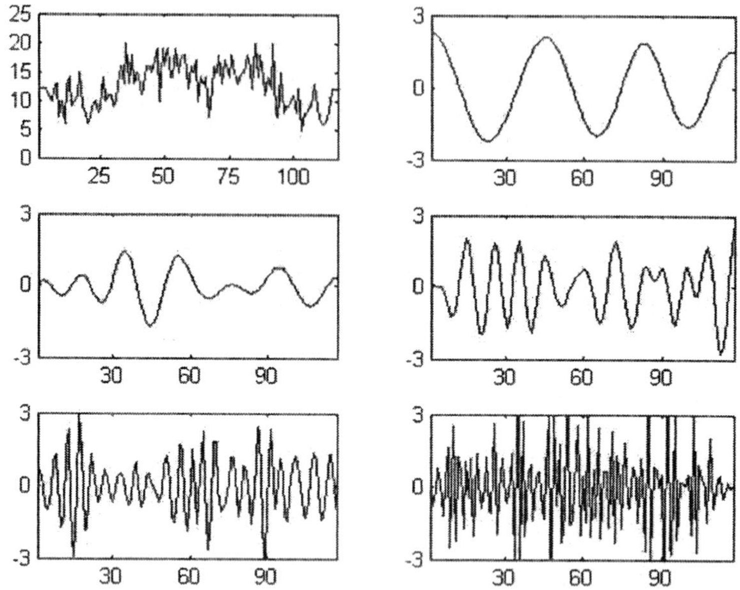

Fig. 2. Wavelet analysis for the frequency of low pressure system (per year). Top left is the original time series and the rest is its different harmonics

Figure 1 represents time series of total number of low pressure systems on an annual basis (5 year moving average is also indicated). The time series of low pressure system indicates predominant multi-decadal variability. To understand the various periodicities in the number of system per year for the 117 years, we subjected the data to wavelet analysis and identified different types of periodicities (Figure 2). The dominant periodicity is the multi-decadal variability with a periodicity of 40 year. This periodicity can be linked with the Atlantic multi-decadal oscillation of about 60-70 years. The amplitude of this periodicity in the low pressure system is almost equal in its ups and downs of its 40 year periodicity. The other dominant periodicity is the 5.5 years and this periodicity is similar to the periodicity of ENSO cycle. However, the correlation value of the total number of low pressure system in a year with the average nino3.4 SST anomaly for the corresponding year is low and non significant. More detailed studies are needed to prove the relationship between the low pressure system with the ENSO cycle. The other periodicities observed are 18.72 and 12 years.

Figure 3 represents trend of number of low pressure system formed during southwest monsoon season over (a) Arabian Sea, (b) Bay of Bengal and (c) Arabian Sea, Bay of Bengal and land together. The trend analysis provides exactly opposite results over the Arabian Sea and Bay of Bengal. Arabian Sea shows a positive trend whereas Bay of Bengal shows a negative trend. In the recent past, super cyclones were formed over the Arabian Sea during the last week of May in 2007 (Gonu) and 2010 (Phet). In fact, organized convection as part of monsoon onset surge was formed in the normal course during last week of May. These organized convections were intensified into low pressure systems and then cyclones. These extreme weather events are rare in the Arabian Sea. The total number of system (formed over Arabian Sea, Bay of Bengal and land together) shows a decreasing trend. It is to be noted that the number of low pressure system formed over the Arabian Sea is small in comparison with that over the Bay of Bengal. Hence decreasing trend for the low pressure systems over the Arabian Sea, Bay of Bengal and land together is reasonable. The increasing trend of number of low pressure system formed over the Arabian Sea is to be viewed in the climate change scenario.

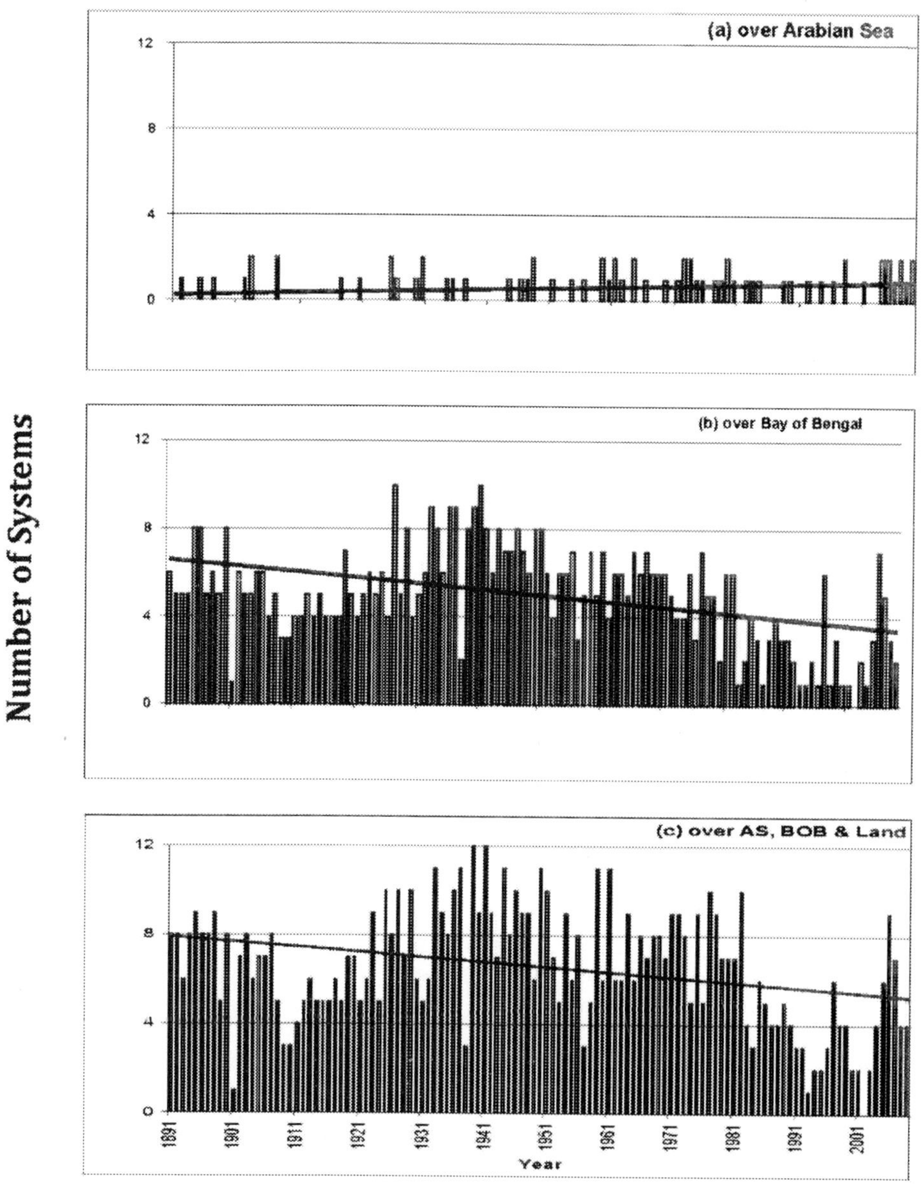

Fig. 3. Trend of number of low pressure system during southwest monsoon season over (a) Arabian Sea, (b) Bay of Bengal and (c) Arabian Sea, Bay of Bengal and Land together

As a next step, the relationship between the annual rainfall and the total number of low pressure system formed in different years was examined. We employed the sliding correlation of 21 year window and found that the annual rainfall and the number of the system have varying relations in the multi-decadal scale (Figure 4). The linear correlation coefficient is positive for the period 1900-1930 and the correlation is reverted. Again it is re-established its positive correlation for the period 1950-1970. Therefore, we can conclude that Indian rainfall has a changing relation with the number low pressure systems formed over the region.

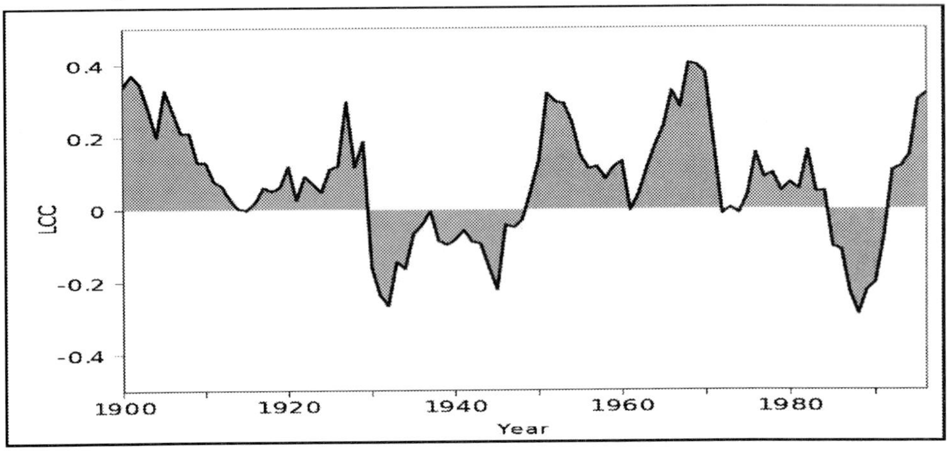

Fig. 4. Linear Correlation Coefficient value between No. of low pressure system and annual rainfall for 21 year sliding window

Conclusion

In this study, we made an analysis on the periodicities embedded on the time series of number of low pressure system formed over the Indian subcontinent, trend analysis on the number of low pressure system formed over the Arabian Sea, Bay of Bengal and total over Arabian Sea, Bay of Bengal and land together during the last 117 years and interrelationship between the number of low pressure system and all India rainfall as sliding correlation of 21 year window. We found that the average number of the low pressure system formed over the region on an annual basis is 12.6 with a standard deviation of 3.7. Maximum and minimum number of the systems were 20 (1925, 1975 and 1982) and 5 (1993) respectively. The dominant periodicity embedded in the time series of number of low pressure system is the multi-decadal variability with a periodicity of 40 year, which can be linked to the Atlantic multi-decadal oscillation of about 60-70 years. The other dominant periodicity is 5.5 years and this periodicity is similar to the periodicity of ENSO cycle. The other periodicities observed are 18.72 and 12 years. The number of low pressure system formed during southwest monsoon over the Arabian Sea shows an increasing trend whereas that over the Bay of Bengal shows a decreasing trend. Total number of system formed over the Indian region also shows a decreasing trend. From the sliding correlation analysis for 21 year window, we found that the annual rainfall and the number of system have varying relations in the multi-decadal time scale. The linear correlation coefficient is positive for 1900-1930 and the correlation is reverted. Again it is re-established its positive correlation for the period 1950-1970. Thus Indian rainfall has a relationship with the number of low pressure systems formed over the region that varies with time.

References

1. Goswami B N, Venugopal V, Sengupta D, Madhusoodanan M S and Prince K Xavier, Increasing trend of Extreme rain events over India in a warming environment, *Science*, 314 (2006) 1442–1445.
2. Feng S and Hu Q, How the North Atlantic Multidecadal Oscillation may have influenced the Indian summer monsoon during the past two millennia, *Geophys. Res. Lett.*, 35 (2008) L01707, doi:10.1029/2007GL032484.
3. Kumar Krishna K, Balaji Rajagopalan, Martin Hoerling, Gary Bates and Mark Cane, Unraveling the Mystery of Indian Monsoon Failure During El Nino, *Science,* 314 (2006) 115–119.
4. Kumar Krishna K, Balaji Rajagopalan and Mark A Cane, On the Weakening Relationship Between the Indian Monsoon and ENSO, *Science,* 284 (1999) 2156 – 2159.
5. Patwardhan S K and Bhalme H N, A study of cyclonic disturbances over India and the adjacent ocean, *Int. J. Climatol.* 21 (2001) 527–534.
6. Hu Q and Feng S, Variation of the North American Summer Monsoon Regimes and the Atlantic Multi decadal Oscillation, *J. of Climate*, 21 (2008) 2371–2378.
7. Krishnan R and Sugi M, Pacific decadal oscillation and variability of the Indian summer monsoon rainfall, *Climate Dynamics.*, 21 (2003) 233–242.
8. Camargo J Suzanna, Adam H Sobel, Anthony G Barnston and Philip J Klotzbach, The influence of natural climate variability on tropical cyclones, and seasonal forecasts of tropical cyclone, Global Perspectives on Tropical Cyclones, from Science to Mitigation, *World Scientific Series on Earth System Science in Asia.*, vol. 4 2nd edition (2009), pp. 325–360.
9. Johnny C L Chan, Decadal variations of intense typhoon occurrence in the western North Pacific, *Proc. R. Soc. A (2008)*, 464 (2007) 249–272 doi:10.1098/rspa. 2007.0183.
10. Johnny C L Chan, Comment on "Changes in Tropical Cyclone Number, Duration, and Intensity in a Warming Environment", *Science*, 311 (2006) 1713b, DOI: 10.1126/science.1121522.
11. Landsea C, Nicholls N, Gray W and Avila L, Downward trends in the frequency of intense Atlantic hurricanes during the past five decades, *Geophysical Research Letters,* 23 (1996) 1697–1700.
12. Webster P J, Holland G J, Curry J A and Chang H R, Changes in tropical cyclone number, duration and intensity in a warming environment, *Science,* 309 (2005) 1844–1846.

13. Johnny C L, Chan Shi J E, Long-term trends and inter annual variability in tropical cyclone activity over the Western North Pacific, *Geophysical Research Letters,* 23 (20) (1996) 2765–2767.

14. Joshi U R, Rajeevan M, Trends in precipitation extremes over India, *National climate centre research report (India Meteorological Department, Pune)*, 3 (2006) 1–25.

The influence of global temperature on the monsoon rainfall over Tamil Nadu

R. Samuel Selvaraj & [1]Raajalakshmi Aditya

Department of Physics, Presidency College, Chennai - 600 005, India
[1]Government Arts & Science College, Tiruttani, India
[1][E mail: rajvi99@gmail.com]

Introduction

Tamil Nadu formerly known as Madras state is the only sub-division of the Indian union which receives more rainfall in the northeast monsoon (October-December) season than in the southwest monsoon (June-September) season. In the previous study based on 80 years rainfall data from 1891, it was found that the southeast and northeast monsoon seasons contribute about 34% and 47% respectively to the mean annual rainfall of this region[1] .Thus the rainfall received in Tamil Nadu during the northeast monsoon season is of great economic value. Major agricultural operations are normally undertaken during that season. It has, however, been noted that the rainfall during northeast monsoon is highly variable. Therefore, if its behavior could be predicted in advance, it would go a long way toward helping the agricultural and industrial activities of the region.

The southwest monsoon is a part of the equatorial westerly, and as such it is humid, unstable and of considerable vertical extent. On the other hand, the northeast monsoon is a part of the northeast trades and is comparatively dry, stable and of lesser vertical extent, about 1 to 2 km. As a result of the fundamental differences in the origin and structure of these two monsoons, Tamil Nadu receives varying amounts of rainfall during the two rainy periods.

Rao[2] observed that in the same year, there is an indication that a good northeast monsoon rainfall over Tamil Nadu is generally associated with deficient or normal southwest monsoon rainfall and vice versa. Raj *et al.*[3] and Raj[4] have identified a few upper air parameters that have predictive potential for northeast monsoon rainfall over Tamil Nadu and the southern region. The annual variability of Indian summer monsoon and northeast monsoon rainfall is determined by external forcings and nonlinear internal dynamics. Surface air temperature is one of the factors that influence monsoon variability. The distribution of surface air temperature over land and sea determines the locations of heat source and sink which in turn affect circulation pattern through thermal and latent heat energy exchange between atmosphere and the surface beneath. A number of studies addressed the relationship between Indian summer monsoon and land and sea surface temperature[5,6].

Rajeevan *et al.*[7] examined the global land surface air temperature anomaly pattern in association with inter annual variability of Indian summer monsoon rainfall. During the last few decades there has been a large number of observation and global circulation model (GCM) sensitive studies relating sea surface temperature (SST) over Indian and pacific oceans with Indian summer monsoon rainfall. Pai[8] discussed the prominent teleconnection pattern in the monthly global gridded surface air temperature anomalies associated with the inter annual variability of Indian summer monsoon rainfall. Balchandran *et al.*[9] suggested that in the correlation coefficient patterns, the positive correlation coefficient regions indicate that when the surface air temperature over these areas are warmer (cooler) than normal, the northeast monsoon is above (below) normal. Similarly, the negative correlation coefficient regions indicate that the surface air temperatures over these areas are cooler (warmer) than normal when the northeast monsoon rainfall is above (below) normal. In a similar manner, the association between the global surface air temperature and Tamil Nadu rainfall for the southwest monsoon, northeast monsoon and annual is examined in the present study.

Materials and methods

For the present study more than 100 years global temperature and Tamil Nadu rainfall data is collected from the Indian meteorological department. The raw data of rainfall for the seasonal months over Tamil Nadu and global temperature are considered. A Spearman rank correlation coefficient[10] is calculated in this paper. This correlation determines whether two variables correlate and also provide a measure of strength of their relationship.

The Spearman rank correlation is a non parametric rank correlation technique used in cases where the data is on the ordinal scale. This test uses the rank order and not the actual values for determining the association between the two sets of values, so it is rank correlation. To study the amount or degree of correlation between two variables say X,Y, we compute the Spearman rank correlation as follows.

$$\rho = 1 - [(6 * \Sigma d^2) / (n^3 - n)]$$

where ρ is the Spearman correlation coefficient. To evaluate this equation the raw data is first ranked with rank 1 to the smallest number and rank n to the nth largest number. Next we obtain a set of rank difference scores as ($d = Rank^X - Rank^y$). The sum of the difference in rank of paired values is zero i.e $\Sigma d = 0$, thus we calculate Σd^2. This rank correlation coefficient varies from –1 to 0 to +1. If two variables of X and Y are equal or tied, they are given the average of the rank values that could have been assigned if no ties recurred. The Spearman rank coefficient becomes less efficient if there are more ties in the data. Table 1 represents the correlation values between the global temperature and Tamil Nadu rainfall for the southwest monsoon, northeast monsoon and annual.

Table 1. Correlation analysis between global temperature and Tamil Nadu rainfall

Year	Length	Southwest monsoon	Northeast monsoon	Annual rainfall
1901-1911	11	0.3022	0.2272	0.2772
1912-1922	11	0.2181	0.2863	0.5704
1923-1933	11	-0.1768	0.1318	0.1272
1934-1944	11	0.5920	0.1250	0.0477
1945-1955	11	0.0201	0.0224	0.1019
1956-1966	11	-0.3136	-0.5227	-0.2409
1967-1977	11	0.1522	0.2022	0.3522
1978-1988	11	0.4409	-0.3727	-0.15
1989-2000	12	0.4000	-0.1953	0.1066

Results and discussion

From Table 1 the correlation values give the correspondence between the global temperature and the Tamil Nadu rainfall. The result of the table show both positive and negative correlation, but over a longer duration of the global cycle there has been a good influence of the global temperature on the rainfall. The trend in the values may be due to the uncertainties present in the rainfall data for 100 years. The input is subjected to uncertainty, as a result of measurement errors and systematic errors due to the random nature of rainfall. This precipitation pattern may be influenced by the irregular topography. The large variability in altitude, slope and aspects may increase variability by means of processes such as rain shading and strong winds. Irrespective of the signs there is a significant relationship between the global temperature and the rainfall occurrences. The sign of the correlations merely indicates the strength of the relevance between global temperature and the rainfall. Thus a weak global temperature will give heavier rainfall and a strong global temperature will give lower rainfall. The raw data graph in Fig. 1 (a), (b) and (C) shows that the peak for global temperature is high when the corresponding peak for the rainfall is low showing the global temperature and the rainfall activity are inversely proportional. The correlation is plotted against the global temperature and southwest monsoon in Fig. 2(a), northeast monsoon in Fig. 2(b) and annual rainfall in Fig. 2(c). This gives more emphasis on the inverse correlation between the global temperature and the rainfall occurrences.

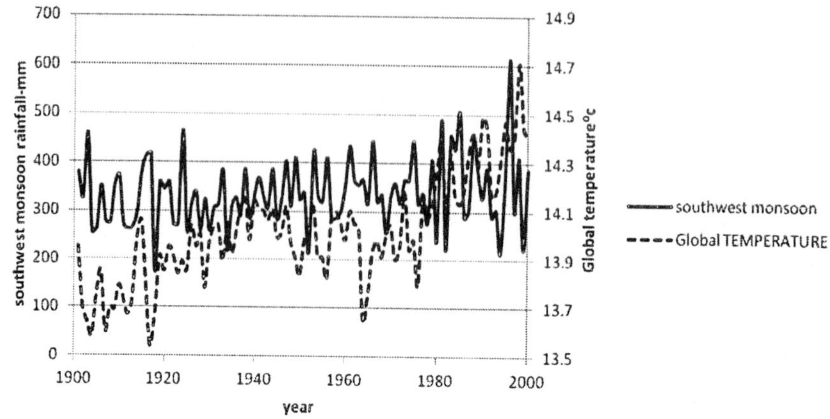

Fig. 1(a). Relation between Global temperature and southwest rainfall of Tamil Nadu.

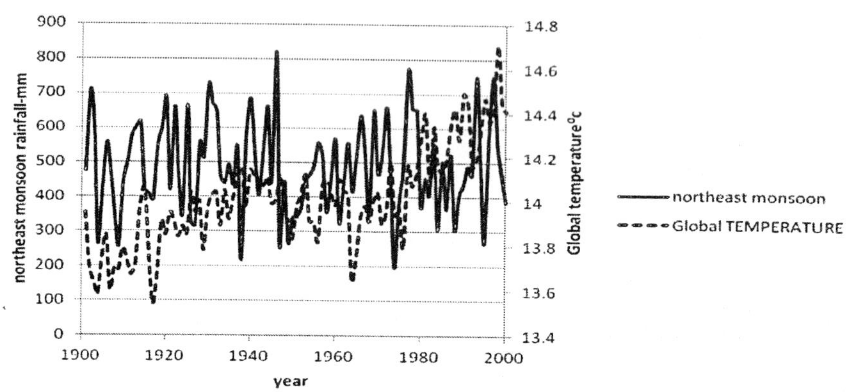

Fig. 1(b). Relation between Global temperature and Northeast rainfall of Tamil Nadu.

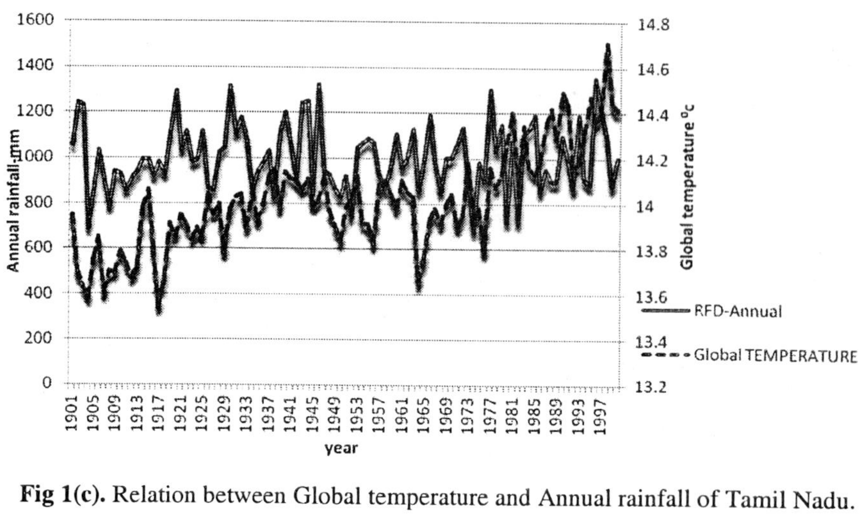

Fig 1(c). Relation between Global temperature and Annual rainfall of Tamil Nadu.

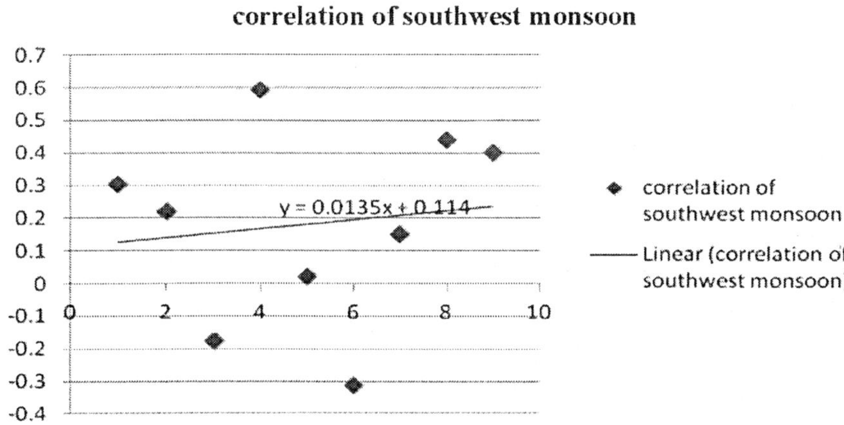

Fig 2(a). correlation between Global temperature and southwest of Tamil Nadu.

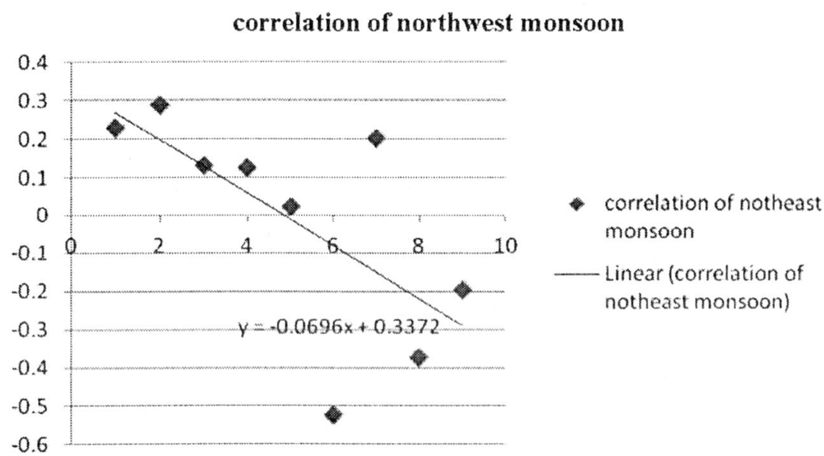

Fig 2(b). correlation between Global temperature and northeast monsoon of Tamil Nadu.

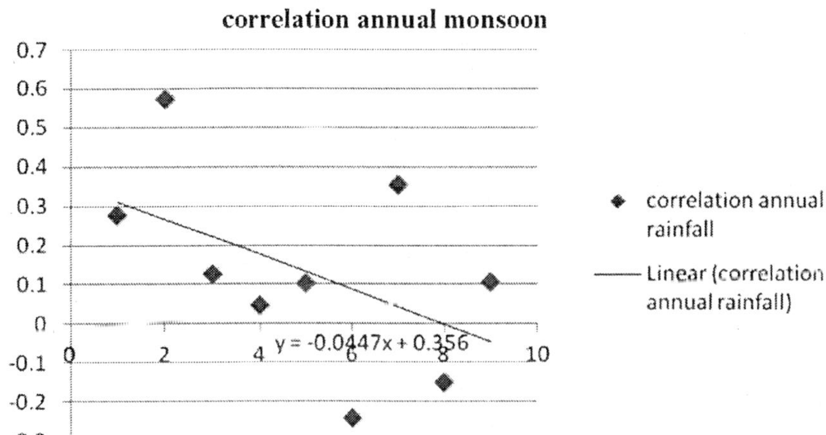

Fig 2(c). correlation between Global temperature and Annual rainfall of Tamil Nadu.

Conclusion

From the correlation table one can infer that for a moderate to higher significance the global temperature is related to Tamil Nadu rainfall. The overall trend of the data shows that the rainfall activity is high for low global temperature and low rainfall activity for higher global temperature occurrences. The physical reason for the above said statement is that the nucleation of water molecules in the atmosphere is brought about by the cosmic rays, when the global temperature increases it will create a magnetic flare which will hinder the entry of the cosmic rays in to the atmosphere. Thus the aggregation of water molecules is not possible in the atmosphere and the rainfall is suppressed.

References

1. Dhar O N, Rakhecha P R and Mandal B N, Some facts about Indian rainfall- A brief appraisal from hydrological consideration, *Indian J. power river Val. Dev.*, (1981) 117–125.
2. Rao P R K, Rainfall of Madras state with special reference to Tamil Nadu and Rayalaseema, Indian meteor. Dept., (1953) 1–60.
3. Raj Y E A, Sen P N and Jamadar S M, Outlook on northeast monsoon rainfall of Tamil Nadu, Mausam (1993) 19–22.
4. Raj Y E A, A scheme for advanced prediction of northeast monsoon rainfall of Tamil Nadu, Mausam (1998) 247–254.
5. Sikka D R; Some aspects of the large scale fluctuations of summer monsoon rainfall over India in relation to fluctuations in the planetary and regional scale circulation parameters, Indian Academic Science, (1980) (Earth Planet Science); vol 89, 179–195.
6. Verma R K, Subramanian K, Dugam S S, Interannual and long term variability of the summer monsoon and its link with northern hemisphere surface air temperature, Indian Academic science (Earth Planet Science) (1985); vol 94, 187–198.
7. Rajeevan M, Pai D S, Thapliyal V, Spatial and temporal relationship between global surface air temperature anomalies and Indian summer monsoon, Meteorol. *Atmospheric Physics* (1998); 157–171.
8. Pai D S, Tele connections of Indian summer monsoon with global surface air temperature anomalies; Mausam (2003); vol 54(2), 407–418.
9. Balachandran S, Asokan R, Sridharan S, Global surface temperature in relation to Northeast monsoon rainfall over Tamil Nadu; *J. Earth. syst. Sci.,* (2006); vol 115, 349–362.
10. Press W H, Teukolsky Vetterling W T and Flannery B P, Numerical recipes in C, Cambridge (1992) 640.

Treatment of acid mine water for disposal in sea

Alka Pradhan[1] *& Jitendra P. Deshmukh*[2]*

[1][Motilal Vigyan Mahavidyalaya, Bhopal],
[2][Mahatama Gandhi Labour Institute, Ahmedabad]*
[E- Mail: jitenddeshmukh@yahoo.com]

Introduction

In 1968, Pennsylvania instituted strict effluent discharge limitations on mine operations. Many mining companies used chemical treatment methods to meet these new effluent limits. In these chemical treatment systems, the acidity is buffered by the addition of alkaline chemicals such as calcium carbonate, sodium hydroxide, sodium bicarbonate or anhydrous ammonia. These chemicals raise the pH to acceptable levels and decrease the solubility of dissolved metals. Precipitates form that are settled from the solution. But these chemicals are expensive and the treatment system requires additional costs associated with operation and maintenance as well as the disposal of metal-laden sludges.

Like wise power plants generates fly ash which contains the oxides of calcium and hence alkalinity. Till date plenty of research work is carried out to know the possible neutralization of acid mine water with the fly ash as it will be economical and eliminate both the problem of Acid Mine Water and Fly Ash treatment disposal. Economic consideration of treatment cost is very important to establish the treatment plant or system. As most of the power plants are situated near to the lignite mine due to the availability of lignite if the fly ash available from these plants can be utilized for treatment of acid mine water of that mine the cost tremendously comes down.

Water flowing from or caused by surface mining, deep mining or coal refuse piles that is typically highly acidic with elevated levels of dissolved metals. The formation of AMD is primarily a function of the geology, hydrology and mining technology employed for the mine site. AMD is formed by a series of complex geo-chemical and microbial reactions that occur when water comes in contact with pyrite (iron disulfide minerals) in coal, refuse or the overburden of a mine operation. The resulting water is usually high in acidity and dissolved metals. The metals stay dissolved in solution until the pH raises to a level where precipitation occurs. Solubility charts for the various metals show the pH at which precipitation begins and the pH at which maximum insolubility occurs.

Basic AMD chemistry

There are four commonly accepted chemical reactions that represent the chemistry of pyrite weathering to form AMD. An overall summary reaction is as follows:

$$4\ FeS_2 + 15\ O_2 + 14\ H_2O \rightarrow 4\ Fe(OH)_3^{\ -} + 8\ H_2SO_4$$

Pyrite + Oxygen + Water → "Ferric Hydroxide" + Sulfuric Acid

The first reaction in the weathering of pyrite includes the oxidation of pyrite by oxygen. Sulfur is oxidized to sulfate and ferrous iron is released. This reaction generates two moles of acidity for each mole of pyrite oxidized.

$$2\ FeS_2 + 7\ O_2 + 2\ H_2O \rightarrow 2\ Fe^{2+} + 4\ SO_4^{2-} + 4\ H^+$$

Pyrite + Oxygen + Water → Ferrous Iron + Sulfate + Acidity

The second reaction involves the conversion of ferrous iron to ferric iron. The conversion of ferrous iron to ferric iron consumes one mole of acidity. Certain bacteria increase the rate of oxidation from ferrous to ferric iron. This reaction rate is pH dependant with the reaction proceeding slowly under acidic conditions (pH 2-3) with no bacteria present and several orders of magnitude faster at pH values near 5. This reaction is refered to as the "rate determining step" in the overall acid-generating sequence.

$$4\ Fe^{2+} + O_2 + 4\ H^+ \rightarrow 4\ Fe^{3+} + 2\ H_2O$$

Ferrous Iron + Oxygen + Acidity → Ferric Iron + Water

The third reaction which may occur is the hydrolysis of iron. Hydrolysis is a reaction which splits the water molecule. Three moles of acidity are generated as a byproduct. Many metals are capable of undergoing hydrolysis. The formation of ferric hydroxide precipitate (solid) is pH dependant. Solids form if the pH is above about 3.5 but below pH 3.5 little or no solids will precipitate.

$$4\ Fe^{3+} + 12\ H_2O \rightarrow 4\ Fe(OH)_3^{\ -} + 12\ H^+$$

Ferric Iron + Water → Ferric Hydroxide + Acidity

The fourth reaction is the oxidation of additional pyrite by ferric iron. The ferric iron is generated in reaction steps 1 and 2. This is the cyclic and self propagating part of the overall reaction and takes place very rapidly and continues until either ferric iron or pyrite is depleted. Note that in this reaction iron is the oxidizing agent, not oxygen.

$$FeS_2 + 14\ Fe^{3+} + 8\ H_2O \rightarrow 15\ Fe^{2+} + 2\ SO_4^{2-} + 16\ H^+$$

Pyrite + Ferric Iron + Water → Ferrous Iron + Sulfate + Acidity

Materials and methods

Experimental description

Several techniques have been developed for the treatment of acid mine drainage which includes lime neutralization, biological sulphate removal and ferrite process etc. Power plants utilizing lignite/coal are also constantly looking for the environmentally sound disposal of fly ash the present available methods are land fill, use in bricks manufacturing or cement plants and construction sites.

Recent studies shows that the treatment of acid mine water is possible with fly ash moreover it becomes more economically viable when the power plant and lignite or coal mine are situated side by side.

The five different samples of fly ash and acid mine water are collected and analyzed in the laboratory to establish the content and characteristic. The parameters are analysed as per the Indian Standard procedure.

Results and discussion

Fly ash

S. No.	Parameter	Result
01.	SiO_2 %	22.74
02.	R_2O_3 %	37.60
03.	Fe_2O_3 %	13.33
04.	Al_2O_3 % (by difference)	17.80
05.	TiO_3 %	5.44
06.	CaO %	10.52
07.	MgO %	2.80
08.	SO_3 %	8.17
09.	Na_2O_3 (ppm)	173
10.	K_2O (ppm)	15.34
11.	Chloride %	1.65
12.	P_2O_5 %	0.046
13.	MnO %	0.98

Characteristics of acid mine water

S. No.	Parameter	Result
01.	pH	2.5
02.	Electrical Conductivity (ms/cm)	10.83
03.	Acidity (mg/l) $CaCO_3$	4005
04.	Total Dissolved Solids (mg/l)	12,000
05.	Sodium (mg/l)	324
06.	Mg (mg/l)	1003
07.	Al (mg/l)	800
08.	Fe (mg/l)	4550

The batch neutralization experiments were carried out by stirring a predetermined mixture of acid mine water and fly ash at room temperature and pressure to obtain a specific fly ash to acid mine water ratio.

The experiment was carried out in each case for 30 minutes and the neutralization process was monitored continuously by the on line digital pH meter. At acid mine water to fly ash ratio of 3:1 and 30 mintes reactivity time the pH of the acid mine water neutralized to 6.6. The time vs pH change is as follows

Conclusion

From the above work it is observed that considerable neutralization of Acid Mine Water can be achieved with the reaction of fly ash containing CaO. The results are encouraging and shows that the pH of acid mine water is elevated from 2.5 to 6.6 for an acid mine water to fly ash ratio of 3:1 within 30 minutes. The other observation is that the amount of fly ash required is 4-5 times higher than the lime requirement therefore, more fly ash can be disposed off.

Acknowledgment

Authors sincerely acknowledge the support and guidance provided by the Chemical Department of Motilal Vigyan Mahavidyalaya Bhopal and Gujarat Mineral Development Corporation Ltd., Ahmedabad.

Alginic acid yield, biochemical composition and biomass in brown seaweed species of Cape Monze shore, Karachi

*Fozia Khan, Qaisar Abbas & Rashida Qari**

Institute of Marine Sciences, University of Karachi, Karachi-75270, Pakistan
[E-mail: rqari2002@yahoo.com]

Introduction

The present uses of the seaweeds are as human food, cosmetics, and fertilizers and for the extraction of industrial gums and chemicals. They have the potential to be used as a source of long and short chain chemicals with medicinal and industrial uses. The Phaeophyceae or brown algae is a large group of filamentous and macroscopic plants, have practical applications as a source of structurally and functionally unique polysaccharides, alginic acid and their salts i.e., Alginates (Shevchenko *et al.*, 2007).

Karachi coast has a large diversity of marine algae, approximately 177 genera and 475 species of marine plants (Shameel and Tanaka, 1992), with great abundance (Qari, 2002; Qari and Qasim, 1994 and Qari and Qasim, 1988). A number of workers carried out research on seaweed chemical constituents and natural products (Qari and Siddiqui, 2010; 2005; 2001 and 1993; Qari and Qasim, 1993; Qari, 1988 and Shameel, 1987). Information on the brown seaweeds polysaccharides alginic acid from Karachi Pakistan is scarce as compared to other countries. However, recently Khan and Qari (2009) studied seasonal variation in water soluble polysaccharide alginic acid extracted from *sargassum boveanum* along the different shores of Karachi coast.

Despite the fact that there are twenty five (25) genera and sixty two species of Phaeophyceae are occurred on the coast of Karachi and most of the species are the perennial species (Qari, 2002; Shameel and Tanaka, 1992; and Qari and Qasim, 1994, 1988). The present study provided useful information for the exploitation of most of the species of brown seaweed had better and high quantity of alginic acid content and biomass that make them valuable as commercial species.

Material and methods

Cape Monze is situated at 24°50′ N 66°42.5′ E on the South Western coast of Arabian Sea. It is between the mouth of the Hub River and the promontory of

Gadani and one of the most beautiful shores. It is still untouched and undeveloped with crystal clear water and sandy beach, as a result it still consist of undisturbed forms of flora and fauna. The jutting rocks lessen the force of the waves thus protecting the backwater beach and along with it the existing life. The Cape Monze shore starts with pale yellow colour rocks at its front, the main boulder like forms, is exposed to the surface and slopes steep towards the sea. The sides of these rock formations are thickly covered with the biota especially rich in seaweed.

Seaweeds were collected from Cape Monze shore of Karachi coast in a month of winter (February). The sampling method of Chapman (1964) was followed for estimating the seaweed biomass. For biochemical estimation seaweed samples were dried at 70°C for 24 hours and were ground in fine homogenized powder. The extraction of alginic acid method is used described by Haug (1964). The determination of purity and the physical properties i.e., relative density and viscosity of alginic acid the methods of Whyte *et al.* (1981) and Whyte (1988) were followed. Moisture and ash contents were estimated by standard method of A.O.A.C. (1990). The carbohydrate was estimated by the phenol-sulphuric acid method reported by Dubois *et al.* (1956).

Results and discussion

Total eleven species belonging to eight genera were collected from Cape Monze of Karachi coast in period of February. The data presented in Table 1 show the mean values with standard deviation of the alginic acid and its physical properties in all species of seaweed. Each reported value of alginic acid yield is the mean of three observations expressed in per cent and its physical properties relative density and viscosity are expressed in units and centipoises respectively. The data reveals high variability in the content of alginic acid and its physical properties in between seaweed species.

The high alginic acid content was found in *Stoechospermum marginatum* (19.68 ± 0.17%) and low in *Colpomenia sinuosa* (10.54 ± 0.37%) as compared to other studied species of seaweeds (Table 1). In between Sargassum species high content of alginic acid was found in *Sargassum boveanum* (16.35 ± 0.29%) as compared to *S. lanceolatum* (13.61±0.06%) and *S. swartzii* (15.39 ± 0.45%). *Spatoglossum variable* has almost the same alginic acid content (16.85 ± 0.1%) like the *Sargassum boveanum* (Table 1).

The ranges of relative density of the alginic acid extracted from different eleven species of seaweeds were almost same (0.77-0.98 unit) with the mean value of 0.898 units (Table 1). The range of viscosity, physical property of alginic acid was 76.87-111.6 cp. The high viscosity of alginic acid was found in extract of alginic acid of *Padina tetrastromatica* (111.6 ± 1.06 cp) as compared to other studied species of seaweeds (Table 1).

Table 1. Alginic acid (%) and its physical properties of brown seaweeds from Cape Monze of Karachi coast.

S. No.	Name of species	Alginic acid	Relative density	Viscosity
1.	*Colpomenia sinuosa*	10.54±0.37	0.98±0.01	95.92±0.56
2.	*Cystoseira indica*	11.52±0.48	0.89±0.01	100.2±0.60
3.	*Dictyota dichotoma*	14.61±0.07	0.87±0.01	88.2±0.74
4.	*Iyengeria stellata*	12.34±0.32	0.86±0.01	100.2±1.16
5.	*Padina pavonia*	14.96±0.69	0.97±0.01	102.2±0.69
6.	*P. tetrastromatica*	14.07±0.25	0.77±0.01	111.6±1.06
7.	*Sargassum boveanum*	16.35±0.29	0.97±0.01	76.87±0.01
8.	*S. lanceolatum*	13.61±0.06	0.89±0.01	95.7±1.15
9.	*S. swartzii*	15.39±0.45	0.88±0.01	88.69±0.95
10.	*Spatoglossum variable*	16.85±0.1	0.91±0.01	97.74±0.90
11.	*Stoechospermum marginatum*	19.68±0.17	0.8±0.01	104.0±0.81

The range of carbohydrates content was 30.85-38.41%. The highest carbohydrates content was found in *Spatoglossum variable* 38.41 ± 1.24 as compared to other studied species. The high ash content was found in *Stoechospermum marginatum* (24.55 ± 0.55%) as compared to other ten studied species of seaweeds (Table 2). In present work all species abundantly found on the coast of Cape Monze, Karachi coast (220-670 gm^{-2}). It is also noted that Sargassum species (*S. boveanum, S. lanceolatum* and *S. swartzii*) are dominant with greater biomass (570-670 gm^{-2}) (Table 2).

Table 2. Biochemical composition (%) and biomass (gm^{-2}) in brown seaweeds of Cape Monze, Karachi coast.

S. No.	Name of species	Carbohydrate	Ash	Moisture	Biomass
1.	*Colpomenia sinuosa*	35.35±5.002	22.01±1.70	2.94±0.44	240±0
2.	*Cystoseira indica*	36.67±1.815	18.14±0.52	3.20±0.74	475±0
3.	*Dictyota dichotoma*	30.85±2.42	21.30±0.935	6.27±0.59	350±0
4.	*Iyengeria stellata*	31.35±4.61	13.26±0.27	3.58±0.05	230±0
5.	*Padina pavonia*	32.36±3.43	20.66±0.56	6.28±0.15	340±0
6.	*P. tetrastromatica*	33.01±3.108	21.47±0.76	5.32±0.16	345±0
7.	*Sargassum boveanum*	31.12±1.13	17.47±1.04	4.08±1.04	570±0
8.	*S. lanceolatum*	32.72±0.91	17.91±0.38	4.32±0.32	670±0
9.	*S. swartzii*	37.82±1.06	16.75±0.53	3.61±0.18	660±0
10.	*Spatoglossum variable*	38.41±1.24	21.21±0.50	4.03±0.38	250±0
11.	*Stoechospermum marginatum*	32.54±2.77	24.55±0.55	2.59±0.58	220±0

The results of analysis of variance (ANOVA) showed that there were no any significant variations found in between alginic acid, biochemical constituent and biomass of different eleven studied species of seaweed. There was also no particular correlation found in between alginic acid, carbohydrate and biomass of studied species of seaweed (Table 3).

Table 3. Correlations (Pearson) between alginic acid, biochemical constituent and biomass of different species of seaweed.

	Carbohydrate	**Ash**	**Moisture**	**Biomass**
Ash	0.051			
Moisture	-0.412	0.093		
Biomass	0.106	-0.441	0.054	
Alginic acid	-0.087	0.405	0.013	-0.047

The results for yield of alginic acid extracted from species of brown seaweeds obtained during the study were similar to the values reported by Arvizu *et al.* (2007) from Baja California Sur, Mexico, Jothisaraswathi *et al.* (2006) from Gulf of Mannar, India and Chandia *et al.* (2004) from South of Chile. Recently Khan and Qari (2009) have also given the same results of alginic acid content in brown seaweeds *S. boveanum* that was collected from the four different shores (Hawks Bay, Buleji, Paradise Point and Manora) of Karachi coast. The viscosity of alginic acid was also similar to the values reported by Khan and Qari (2009) and Jothisaraswathi *et al.* (2003). The results of carbohydrate and ash were recorded in the present study have much resemblance with the results of Chakraborty and Santra (2008) and Hossain *et al.* (2003).

It is noted that high concentrations of carbohydrate and alginic acid were due to immature or young stage of plant (Azad *et al.*, 2007). The present study provide useful information for the exploitation of most of the brown seaweed species that had better and high quantity of alginic acid content and biomass that make them valuable as commercial species. So it is concluded that brown seaweed especially *Stoechospermum marginatum* and all studied *Sargassum* species (*S. boveanum, S. lanceolatum* and *S. swartzii*) and *Spatoglossum variable* found on the coast of Karachi can be used as a source of alginic acid in different kinds of food, medicines, textiles, paper, cosmetics and fertilizer industries. It is also noted more research is necessary to evaluate the nutritional value of seaweed as food ingredients in Pakistan because still seaweeds are not in use as food in Pakistan.

References

1. A O A C, 1990. Official methods of analysis. 15th edition, *Association Official Analytical Chemists*. Arlington, Virginia. 684 pp.
2. Arvizu D L, Rodriguez Y E, Hernandez G and Murillo J I, 2007. Chemical constituents of *Eisenia arborea* Areschoug from Baja California Sur, Mexico. *Investig. Mar.*, 35(2): 63-69.

3. Azad B, Alauddin M, Islam, M S, Hoque M R and Chowdhury Z, 2007. Study on biochemical composition of brown seaweeds collected from Saint Martin's Island of Bangladesh. *European Journal of Scientific Research*, 17(1): 97-105.

4. Chakraborty S and Santra S C, 2008. Biochemical composition of eight benthic algae collected from Sunderban. *Indian Journal of Marine Sciences*, 37(3): 329-332.

5. Chandia N P, Matsuhiro B, Mejias E and Moenne A, 2004. Alginic acid in *lessonia vadosa*: Partial hydrolysis and elicitor properties of the polymannuronic acid fraction. *Journal of Applied Phycology*, 16: 127-133.

6. Chapman V J, 1964. Coastal vegetation. Pergaman Press, Oxford.

7. Dubios M, Gilles K A, Hamilton J K, Rebers P A and Smith F, 1956. Colori-metric method for determination of sugars. *Anal. Chem.* 28: 350 pp.

8. Haug A, 1964. Composition and properties of alginates. *Rep. Norw. Inst. Seaweed Res.* (30): 123pp.

9. Hossain Z, Kurihara H and Takahashi K, 2003. Biochemical composition and lipid compositional properties of brown Alga *Sargassum horneri. Pakistan Journal of Biological Sciences* 6(17): 1497-1500.

10. Jothisaraswathi S, Babu B and Rengasamy R, 2003. Seasonal studies on alginate and its biochemical composition I: *Sargassum polycystum.* (Fucales), Phaeophyceae. *Phycological Research*, 51: 240-243.

11. Jothisaraswathi S, Babu B and Rengasamy R, 2006. Seasonal studies on alginate and its composition II: *Turbinaria conoides* (J.Ag.) Kutz. (Fucales, Phaeophyceae). *J. Appl. Phycol.*, 18: 161-166.

12. Khan F and Qari R, 2009. Seasonal; variation in water soluble polysaccharides alginic acid extracted from *Sargassum boveanum* J. Agardh (Phaeophyta, Sargassaceae) along the different shore of Karachi coast, *Pakistan. Int. J. Biol. Biotech.*, 6(4): 277-281

13. Qari R, 1988. Seasonal changes in Biochemical composition of seaweed from Karachi coast of Pakistan. *Pakistan J. Sci. Ind. Res.,* 31: 94-96.

14. Qari R, 2002. Studies of biodeposited trace metals and minerals in marine algae from Karachi coast. *PhD thesis*, University of Karachi. 454 pp.

15. Qari R and Qasim R, 1988. Seasonal change in the standing crop of intertidal seaweeds from the Karachi coast. *In*: Proc. *Marine Science of the Arabian Sea.* Thompson M. F. and Tirmizi N. M., eds. American institutes of biological sciences. Washington D. C., 449-456.

16. Qari R and Qasim R, 1993. Biochemical constituents of seaweed from Karachi coast. *Indian J. Mar. Sci.*, 22: 229-231.

17. Qari R, and Qasim R, 1994. Seasonal change in the standing crop of intertidal seaweeds from Manora coast, Karachi. *In*: Proc. *Nat. Sem. Fish Policy and Plan.*, Majid A, Khan M Y, Moazzam M, Ahmed J, eds. Marine Fisheries Department Karachi, 279-286.

18. Qari R and Siddiqui S A, 2001. A seasonal study for establishing *Ulva fasciat* (Chlorophyta) as a biological monitor for pollution at Buleji and Paradise Point, near Karachi. Pakistan *J. Mar. Biology*, 7 (1 and 2): 261-280.

19. Qari R and Siddiqui S A, 1993. Biochemical composition and yield of agar from the *Gracilaria corticata* of Karachi. *Pakistan J. Mar. Biology*, 2: 77-81.

20. Qari R and Siddiqui S A, 2005. Variations of heavy metals in green seaweeds from Karachi coast of Pakistan. *Pakistan J. Sci. Ind. Res.*, 48(3): 195-201.

21. Qasim R, 1981. Biochemical studies on some seaweed from Karachi coast. *Karachi, Univ. J. Sci.*, 9: 105-111.

22. Qasim R and Barkati S, 1985. Ascorbic acid and dehydroascorbic acid contents of arine algal species from Karachi. *Pakistan J. Sci. Ind. Res.*, 28: 129-133.

23. Qari R and Siddiqui S A, 2010. A comparative study of heavy metal concentrations in red seaweeds from different coastal areas of Karachi, Arabian Sea. IJMS 39(1): 27-42. 8.

24. Shameel M, 1987. A preliminary survey of seaweed from the coast of Lasbella, Pakistan. *Bot. Mar.*, 30: 511-515.

25. Shameel M and Tanaka J, 1992. A preliminary checklist of marine algae from the coast and inshore waters of Pakistan. *In: Cryptogamic Flora of Pakistan*. Nakaike, T. and Malik, S., eds. 1: Nat. Sci. Mus., Tokyo, 1-64.

26. Shevchenko N M, Anastyuk S D, Gerasimenko N I, Dmitrenok P S, Isakov V V and Zvyagintseva T N, 2007. Polyssacharides and lipid composition of the brown seaweed *Laminaria gurjanovea. Russian Journal of Bioorganic Chemistry*, 33(1): 88-98.

27. Whyte J N C, Engler J R, Saunders R G and Landsay J C, 1981. Seasonal variations in the Biomass, Quantity and Quality of Agar, from the Reproductive and Vegetative Stages of *Gracilaria* (*verrrucosa* type). *Botanica Marina*, 24: 493-501.

28. Whyte J N, 1988. Extraction of alginic acid from brown seaweed *In*: *Experimental Phycology – A Laboratory Mannual*. Lobban, C. S., Chapman, D. J. and Kremer B. P., eds. Cambridge, University Press, Cambridge. 168-183.

Establishing India's botanic gardens network as a clearing house mechanism in times of rapid global change

Sudershan Kumar, Uma Supyal, Leena Wahi, Abhilasha Agnihotri, Bajrang Singh & A.K. Goel

CSIR, National Botanical Research Institute, Rana Pratap Marg, Lucknow – 226 001 (UP), India
[E-Mail: sudershank@hotmail.com]

Introduction

Plants are important sources of oxygen and alleviating, carbon dioxide concentration in the atmosphere, food and drinks, basis of most food chains, clothing, shelter, prevention of soil erosion, habitats for many other organisms, drugs, building materials, paper, fuels, fibres, perfumes, cosmetics, soaps, sports equipment, recreation areas and jobs (plant agriculture is a major employer). Plants form a major role in healthcare. Traditional Chinese medicine alone uses over 5,000 plant species and traditional medicines in India are based on 7,000 different plants. Today plants are used more than ever in medicines and pharmaceutical companies are always searching for plants with new useful properties. We owe an awful lot to plants, it is very important that we stem the loss of valuable species now[1]. The earth's climate is changing more rapidly than at any other time in human history. It is now well established that human-induced climate change may result in large-scale plant diversity loss. In particular, climate change could cause dramatic shifts in species distributions and species extinctions, particularly across fragmented or vulnerable ecosystems. About 1.75 million species of organisms in world, India accounts for 7-8% consisting of ca 127,000 species (i.e. 47,000 plants and 81,000 animals) but occupies only 2.4% of the land of the world total. This great diversity is due to varied climate and topographies and due to this India is one of the 12 mega diversity countries and also one of the 12 primary centers of origin of cultivated plants and is rich in agricultural biodiversity.

For four centuries, botanic gardens have led research has generated much of our knowledge about plants and they are important sites of plant conservation because they maintain plants away from their natural habitat (*ex situ* conservation) and have a significant role in education. The Convention on Biological Diversity (CBD), the International Agenda for Botanic Gardens in Conservation and the Global Strategy for Plant Conservation (GSPC) have helped to define and clarify the roles of botanic gardens internationally. Modern botanical gardens are significant agencies for natural history collection, plant *ex situ* conservation, taxonomic research, horticultural and economic botany, public education and natural history appreciation. A key role among the many that have been identified for botanic gardens in a 21st Century context is plant conservation. As a response to the significant global challenge that climate change presents for plant conservation, it is essential to establish a long term monitoring program of plant responses to environmental change using species diversity of India's botanic gardens.

Botanic gardens are important centres for conservation, recreation and education. There are over 2,500 botanic gardens in the world, which maintain the largest collection of plant species outside nature. In the wild, as many as 60,000 of these plant species may be threatened with genetic impoverishment or even be at the verge of extinction within the next 30-40 years. Threats include such factors as habitat loss and fragmentation, introduced species, over-exploitation of plant and animal species, pollution of soil, water and atmosphere, global climatic change, industry, agriculture and forestry. Botanic gardens have an obvious and vital role to play in conserving plants but conservation cannot succeed without education. Gardens are uniquely placed to teach people about the importance of plants in our lives and in the global ecosystem. By highlighting the threats that plants and habitats face, gardens can help people look at ways in which biodiversity can be protected[11].

Botanic gardens face both challenges and opportunities in responding to global and in particular, climate change. Firstly, they will have to reassess their accession policies as the growing conditions in the predicted new climate envelopes for some if not many, of the plants currently cultivated will no longer be suitable for them. Secondly, the increased numbers of species put at a risk as a result of changing climatic conditions will force them to refocus and strengthen their conservation policies. Thirdly, botanic gardens will face an unprecedented opportunity to develop their role as introduction centers and play a major role in the assessment of new germplasm both of ornamentals, medicinal as well as other economically important plants. These three changes are interrelated and hence there is an urgent need to focus on research in this area[11].

Never before has such a widely accepted and powerful international framework existed into which the work of botanic gardens in plant conservation can be fitted. Working through a unique partnership of international and national organizations, Parties, other Governments and NGOs, the 'Convention on Biological Diversity' (CBD) has developed Global Strategy for Plant Conservation (GSPC). The most innovative element of the GSPC is the inclusion of 16 outcome-orientated targets, aimed at achieving a series of measurable goals with, following the 10th Conference of the Parties to the CBD in 2010, a new time horizon up to 2020. This is the first time that such targets have been adopted under the Convention, and the success of this approach will be watched with interest as a potential model for other work programmes.

One of the targets of GSPC is to establish and strengthen networks for plant conservation activities at national, regional and international levels. National networks of botanic gardens help gardens work together to share information and coordinate a united role in the country particularly to face challenges of climate change[2].

At present there are over 100 botanic gardens in India holding both Indian and exotic plant species. It is now recognized that Indian Botanic Gardens could be excellent sites for the conservation of indigenous plants, including threatened flora and wild species of potential economic importance.

In 2002 Botanic Gardens Conservation International (BGCI) launched an international plant programme conservation programme funded by HSBC (Investing in Nature (IiN) programme). The programme had operated at the global level, implementing the International Agenda for Botanic Gardens in Conservation as well specific national programmes in Argentina, Brazil, China including Hong Kong, India, Indonesia, Japan, the Middle East, North America (including Canada and the USA) and South-East Asia[3]. In India it was implemented by the Bioinformatics Division of the National Botanical Research Institute under Investing In Nature-India (IIN-India) project, an HSBC funded collaboration between BGCI, UK and NBRI, formed to assist in the work being done by botanic gardens in India[4-8].

The present paper deals with the formation of the first ever network of botanic gardens of India called Indian Botanic Garden Network (IBGN), Indian Botanic Garden Code–Index (IBGC-Index) and a data base of 565 rare, endangered and threatened (RET) species as a resource to tackle challenges of climate change. Further strengthening of IBGN with more value addition and setting up of national goals for meeting the challenges posed by climate changes are also discussed in the present paper.

Materials and methods

For the past several hundred years, botanical gardens are only the place where living plants are maintained other than nature. There are approximately over 100 botanical gardens in India today with approximately 200-250 associated Directors and curators and other staff. For developing the first ever network of Indian botanic gardens a standard proforma was designed and circulated to all botanical gardens of India. Proforma was containing parameters for recording information on Botanical Garden name, type, status, curator name, geographical location, botanic garden activities, research areas, total number of taxa in cultivation etc.

So far, of the 95 botanic gardens about 33 botanic gardens have provided data on RET species growing in their respective garden. These species were checked with authentic source and nomenclature of each species including author name and family names of each species were updated. All species habit data was also collected and added to database. A sample of three species showing editing of data is presented in next page.

Species name as provided by botanic gardens	Editing of data from authentic source		
	Family	Correct species name	Citation
Aponogeton appendiculatus	Aponogetonaceae	*Aponogeton appendiculatus* H. Bruggen	Blumea xvi. 265 (1968).
Aponogeton bruggenii		*Aponogeton bruggenii* S.R. Yadav & R.S. Govekar	Rheedea 4(1): 34 (1994).
Aponogeton satarensis		*Aponogeton satarensis* R. Sundararaghavan, A.R. Kulkarni & S.R. Yadav	Kew Bull. 36(4): 687 (1982).

For developing a database, an IBGN software has been developed using PHP (Hypertext Preprocessor) scripting language version 4.3.7 on Linux OS. The software has three modules viz., I- organization online registration, II- individual member directory (IMD) and III- maintenance module. These modules can be accessed in different forms from internet. The layout and the designing of the screens for these modules have been done using the software namely Microsoft Front Page and Macromedia Dreamweaver. Several data validation checks have been incorporated in software to ensure entry of quality data. This software also allows to develop a database on Indian Botanic Garden Code Index (IBGC-Index) and species diversity of India's botanic gardens. IBGC-index entry for a botanic garden includes botanic garden name, contact information and total number of taxa in cultivation, etc. Each botanic garden is assigned a unique code consisting of three components, the first a two letter state name code indicating state, the second a three letter district code indicating district name and the third brief acronym of botanic garden.

Indian Botanic Garden Resource (IBG-Res) database has been developed using software (namely MySQL version 4.0.20a-nt on Linux OS). The IBG-Res database contains information on botanic gardens of India including botanic garden name, type, status, human resource profile, activities, research areas, and public interaction areas of expertise of its associated staff, its physical location details and so on. It also stores information of individual members (IBGN-IMD) including their personal details, contact details, official details and specialization areas/activities. A privilege and password system is provided for the IBG-Res database. Currently there are about 30 tables and 1700 records in the database.

Indian Botanic Garden species (IBG-Sps) database has also been developed using software (namely MySQL version 4.0.20a-nt on Linux OS). The IBG-Sps database contains information on nomenclature, life form, description, uses and images etc. of species growing in various botanic gardens of India. A privilege and password system is provided for the IBG-Res database. A sample list of 565 RET species is presented in the present paper.

Results and discussions

In tandem with the establishment of the Indian botanic garden network, the following resources are developed:

Indian Botanic Garden Code Index (IBGC-Index) of 95 botanic gardens has been developed and presented in Table 1. IBGC-Index is the first ever effort to establish an electronic sharing of data among botanic gardens of India. IBGC-Index will continuously grow and serve in many fields in biodiversity research of India and hence for effective and speedy communication among Botanic Gardens IBGC-Index has been developed. IBGC-Index provides a unique code to each garden which begins with two-letter code (e.g. AP, UP) to state/union territory and 3 letter district code (e.g. PO-AUR1-ABG) to which it belongs, followed by a short abbreviated name for each botanical garden. IBGC-Index will be immensely helpful to give unique identifier to each accession, which are stored in botanic gardens of India, like various herbaria in India and other parts of the world. IBGC-Index will be made mandatory to follow in future subject to approval of concerned authorities and it will also allow new botanic gardens to assign their codes in a meaningful way and also compatible to policy of IBGN.

Table 1. Indian botanic garden code index (IBGC-Index) showing botanic garden name, state name and total number of taxa under cultivation.

S.No.	Name of Botanical Garden	State	Code	Total Accessions/ Holdings
1.	ANSA – MPCP, TIRUCHY	Tamil Nadu	TN-TIU1-ANSABG	000
2.	AROGYADHAM, Deendayal Research Institute, Botanic Garden, SATNA	Madhya Pradesh	MP-STN1-ADRI	450
3.	Auroville Botanical Garden, AUROVILLE	Pondicherry	PO-AUR1-ABG	260
4.	Ayya Nadar Janaki Ammal College Botanical Garden, VIRUDHUNAGAR	Tamil Nadu	TN-VIR1-ANJACBG	100
5.	Belgaum Integrated Rural Development Society (BIRDS), Botanic Garden, BELGAUM	Karnataka	KA-BEL1-BIRDS-BG	000
6.	Botanic Garden of Indian Republic, NOIDA	Uttar Pradesh	UP-NOI1-BGI	000
7.	Botanic Garden of the Central Drug Research Institute, LUCKNOW	Uttar Pradesh	UP-LKO3-CDRIBG	383
8.	Botanic Garden, Dr. BAMU AURANGABAD	Maharashtra	MA-AUN1-BAMUBG	150
9.	Botanical Garden "Dr H.S. Gour Vishwavidyalaya", SAGAR	Madhya Pradesh	MP-SAG1-HSGVBG	215
10.	Botanical Garden cum Arboretum, Dr. YS Parmar University, BG, SOLAN	Himachal Pradesh	HI-SOL1-BGCA	2000
11.	Botanical Garden of Mysore University, MYSORE	Karnataka	KA-MYS1-MUBG	1042

S.No.	Name of Botanical Garden	State	Code	Total Accessions/ Holdings
12.	Botanical Garden, GKVK, BANGALORE	Karnataka	KA-BNG1-GKVKBG	000
13.	Botanical Garden, Jiwaji University, GWALIOR	Madhya Pradesh	MP-GWA1-JUBG	000
14.	Botanical Garden, North-Eastern Hills University, East Khasi Hills, SHILLONG	Meghalaya	ME-SHI1-NEHUBG	500
15.	Botanical Garden, University of Pune, PUNE	Maharashtra	MA-PUN3-UPBG	000
16.	Botanical Garden,TFRI, JABALPUR	Madhya Pradesh	MP-JAB1-TFRIBG	000
17.	Botany Department, Visvabharati, BIRBHUM	West Bengal	WB-BIR1-BDVB-BG	250
18.	Calcutta Wildlife Society, KOLKATA	West Bengal	WB-KOL3-WLSBG	000
19.	Calicut University Botanical Garden, MALLAPURAM	Kerala	KE-MAL1-CUBG-BG	25
20.	CAZRI Botanic Garden, JODHPUR	Rajasthan	RA-JOD2-CAZRIG	000
21.	CCMB Experimental Farm and Landscape, Botanic Garden, HYDERABAD	Andhra Pradesh	AP-HYD2-CCMB-BG	1668
22.	Central Institute of Medicinal & Aromatic Plants, Botanic Garden, LUCKNOW	Uttar Pradesh	UP-LKO2-CIMAPBG	000
23.	Central of Indian Medical Heritage Botanical Garden, COIMBATORE	Tamil Nadu	TN-COI3-CIMHBG	000
24.	CHAI Herbal Garden, Ranga Reddy, SECUDERABAD	Andhra Pradesh	AP-SEC1-CHAIHG	350
25.	Coimbatore Zoological Park & Conservation Centre, COIMBATORE	Tamil Nadu	TN-COI1-CZPCC	500
26.	Community Agro bio diversity Centre conservation Garden, WAYANAD	Kerala	KE-WAY1-CABCCG	212
27.	Darjeeling Botanic Garden, DARJEELING	West Bengal	WB-DAR1-DBG	000
28.	Desert Botanic Garden (BSI), JODHPUR	Rajasthan	RA-JOD1-BSIBG	000
29.	Dr. H.K. Baruah Regional Botanical Resource, Guwahati, KAMRUP	Assam	AS-KAM1-HKBRG	150

S.No.	Name of Botanical Garden	State	Code	Total Accessions/ Holdings
30.	Dr. Kirti's Botanical Garden, IMPHAL EAST	Manipur	MN-IMPH1-KBG	500
31.	Empress Botanical Gardens, PUNE	Maharashtra	MA-PUN1-EBG	56
32.	Entomology Research Institute, CHENNAI	Tamil Nadu	TN-CHE1-ERIBG	500
33.	Ethno-Medicinal Garden, FRLHT, BANGALORE	Karnataka	KA-BNG2-EMG	10
34.	Experimental Botanic Garden Andaman & Nicobar Circle, BSI DHANIKHERI	Andaman & Nicobar	AN-DHK-EBG	250
35.	Experimental Botanic Garden Arid Zone Circle, JODHPUR	Rajasthan	RA-JOD3-EBG	185
36.	Experimental Botanic Garden Central Circle, ALLAHABAD	Uttar Pradesh	UP-ALL2-EBG	570
37.	Experimental Botanic Garden Eastern Circle (Shillong), BARAPANI	Meghalaya	ME-SHI1-EBG	750
38.	Experimental Botanic Garden Northern Circle, DEHRADUN	Uttarakhand	UK-DEH2-EBG	350
39.	Experimental Botanic Garden Northern Circle, KHIRSU	Uttarakhand	UK-KHI1-EBG	000
40.	Experimental Botanic Garden Northern Circle, PAURI	Uttarakhand	UK-PAU1-EBG	750
41.	Experimental Botanic Garden Sikkim Himalayan Circle, GANGTOK	Sikkim	SK-GAN1-EBG	200
42.	Experimental Botanic Garden Southern Circle, Yercaud, SALEM	Tamil Nadu	TN-SAL1-EBG	1200
43.	Experimental Botanic Garden Western Circle, Mundhwa, PUNE	Maharashtra	MA-PUN2-EBGWC	500
44.	Experimental Botanic Garden, Arunachal Field Station, Sankie View, NEW ITANAGAR	Arunachal Pradesh	AP-ITA-AFSEBG	195
45.	Forest Research Institute, DEHRADUN	Uttarakhand	UK-DEH1-FRIBG	400
46.	G. B. Pant Institute Of Himalayan Environment And Development, ALMORA	Uttarakhand	UK-ALM1-IHED-BG	000
47.	G.B. Pant Botanical Garden, PANTNAGAR	Uttarakhand	UK-PNT1-GBPBG	700

S.No.	Name of Botanical Garden	State	Code	Total Accessions/ Holdings
48.	Government Botanic Gardens, OOTACAMUND	Tamil Nadu	TN-NIG1-GBG	272
49.	Gramodava Research and Training Center, Botanic Garden, TUMKUR	Karnataka	KA-TUM1-GRTC-BG	000
50.	Gujarat University Botanic Garden, AHMEDABAD	Gujarat	GU-AHM1-GUBG	000
51.	Guru Nanak Dev University, AMRITSAR	Punjab	PU-AMR1-GDUBG	000
52.	Gurukul Botanical Sanctuary Foundation, CHENNAI	Tamil Nadu	TN-CHE2-GBS	000
53.	Himalayan Botanic Garden, NAINITAL	Uttarakhand	UT-NAI1-HBG	262
54.	Indian Botanic Garden, KOLKATA	West Bengal	WB-HOW2-IBG	1200
55.	Indian Institute of Horticultural Research (IIHR), BANGALORE	Karnataka	KA-BNG3-IIHRBG	000
56.	Institute of Forest Genetics & Tree Breeding (IFGTB), COIMBATORE	Tamil Nadu	TN-COI2-IFGTBG	190
57.	K.U. Dharwad Botanic Garden, DHARWAD	Karnataka	KA-DHW1-KUBG	000
58.	Kashmir University Botanical Garden, SRI NAGAR	Jammu & Kashmir	JK-SRN1-KUBG	700
59.	Kerala Forest Research Institute, THRISSUR	Kerala	KE-THR1-KFRIBG	150
60.	Kerala University Botanic Garden, THIRUVANANTHAPURAM	Kerala	KE-THV1-KUBG	1500
61.	Kodaikanal Botanic Garden, DINDIGUL	Tamil Nadu	TN-DIN1-KBG	000
62.	Lalbagh Botanical Garden, BANGALORE	Karnataka	KA-BNG5-LBHBG	1854
63.	M. S. Swaminathan Research Foundation, CHENNAI	Tamil Nadu	TN-CHE3-MSSRF	000
64.	Madras Christian College, CHENNAI	Tamil Nadu	TN-CHE4-MCC	000
65.	Madurai Kamraj University Botanic Garden, MADURAI	Tamil Nadu	TN-MDU1-MKUBG	000
66.	Magadh University Botanical Garden, GAYA	Bihar	BI-GAY1-MUBG	1200
67.	Maharaja Sayajirao University Botanical Garden, VADODARA	Gujarat	GU-VAD1-MSUBG	000

S.No.	Name of Botanical Garden	State	Code	Total Accessions/ Holdings
68.	Malabar Botanical Garden, KOZHIKODE	Kerala	KE-KOZ1-MBG	000
69.	Manipur University Botanic Garden, IMPHAL, West Kameng	Manipur	MN-IMPH1-MUBG	000
70.	Medicinal Plant Conservation Sugavanam, PUDUKKOTTAI	Tamil Nadu	TN-PUD1-MPCS	300
71.	Medicinal Plant Garden, Regional Research Institute (Ay.), PUNE	Maharashtra	MA-PUN4-MPGRRI	000
72.	Medicinal Plants Botanic Garden, Ranikhet, ALMORA	Uttarakhand	UT-ALM2-MPBG	000
73.	Medicinal Plants Garden-Regional Research Institute (Ayurveda) CCRAS ITANAGAR, Papumpare	Arunachal Pradesh	AR-ITA1-RRIMPG	000
74.	National Botanical Research Institute, LUCKNOW	Uttar Pradesh	UP-LKO1-NBRIBG	6000
75.	NRCMAP Botanical Garden for Medicinal Plants, ANAND	Gujarat	GU-ANA1-MPBG	000
76.	NTR Botanic Garden, HYDERABAD	Andhra Pradesh	AP-HYD1-NTRBG	000
77.	P.N. Mehra Botanical Garden, CHANDIGARH	Chandigarh	CH-CHD1-PNMBG	1000
78.	PRAGYA, GURGAON	HARYANA	HR-GUR1-PG	000
79.	Rani Chandrami Devi Hospital & Biodiversity Park, VISHAKHAPATNAM	Andhra Pradesh	AP-VIK1-RCDHBP	000
80.	Roxburgh Botanic Garden, ALLAHABAD	Uttar Pradesh	UP-ALL1-RBG	500
81.	Satpuda Botanic Garden - College of Agriculture, NAGPUR	Maharashtra	MA-NAP1-SBG	980
82.	SFRI Arboretum, State Forest Research Institute, JABALPUR	Madhya Pradesh	MP-JAB2-SFR-BG	000
83.	Sri Chamarajendra Park, BANGALORE	Karnataka	KA-BNG4-SCP	7200
84.	Sri Kotla Vijaya Bhaskara Reddy Botanical Garden, HYDERABAD	Andhra Pradesh	AP-HYD3-SKVBRBG	300
85.	State Botanical Garden, KHURDA	Orissa	OR-KHU1-SBG	000
86.	State Forest Research Institute, PAPUM PARE	Arunachal Pradesh	AR-PPP1-SFRIBG	454

S.No.	Name of Botanical Garden	State	Code	Total Accessions/ Holdings
87.	Tagore Botanical Garden, JABALPUR	Madhya Pradesh	MP-JAB3-TGBG	000
88.	The Anglade Institute of Natural History, DINDIGUL	Tamil Nadu	TN- DIN1-AINHBG	360
89.	The Rapinat Herbarium, TIRUCHIRAPALLI	Tamil Nadu	TN-TIR1-TRH	000
90.	Tipi, Orchid Research Centre, WEST KAMENG	Arunachal Pradesh	AR-WEK1-ORDC	000
91.	Tropical Botanic Garden and Research Institute, THIRUVANANTHAPURAM	Kerala	KE-THV2-TBGRIG	7500
92.	University Botanic Garden, BHAGALPUR	Bihar	BI-BHG1-BUBG	2719
93.	University Botanic Garden, SURAT	Gujarat	GU-SUR1-UBG	193
94.	Veermati Jijabai Bhosle Udyan-Zoo, Byculla, MUMBAI	Maharashtra	MA-MUM1-VJBUZ	000
95.	Victoria Memorial Garden, KOLKATA	West Bengal	WB-KOL1-VMG	000

Note: A few botanic gardens have not provided data so far and hence data is not shown against their name and efforts are on to collect data from these gardens and update IBGC-Index.

All botanic gardens of India have been grouped into six zones namely I-North Zone, II-South Zone, III-East Zone, IV-West Zone, V-Central Zone, VI-North East Zone and the number of gardens represented in each garden are represented in Figure 1 and the study and data presented in Table 2 reveal maximum number of 38 (40%) botanic gardens in South Zone, followed by 19 (20%) in North Zone, 14 (15%) in West Zone, 10 (11%) in North East Zone 8 (8%) in East Zone and minimum 6 (6%) gardens in Central Zone.

Table 2. Distribution of botanic garden in six zones of India

Name of zone	Total botanic gardens in a zone	%
South Zone	38	40
North Zone	19	20
West Zone	14	15
North East Zone	10	11
East Zone	8	8
Central Zone	6	6
Total	**95**	

Of the 95 botanic about 33 botanic gardens have provided data on RET species growing in their respective garden. These species were checked with authentic source and nomenclature of each species was updated and author name and family names of each species were added. All species habit data was also collected and added to database and thus an edited RET list of 565 RET species presented in Table 3[9].

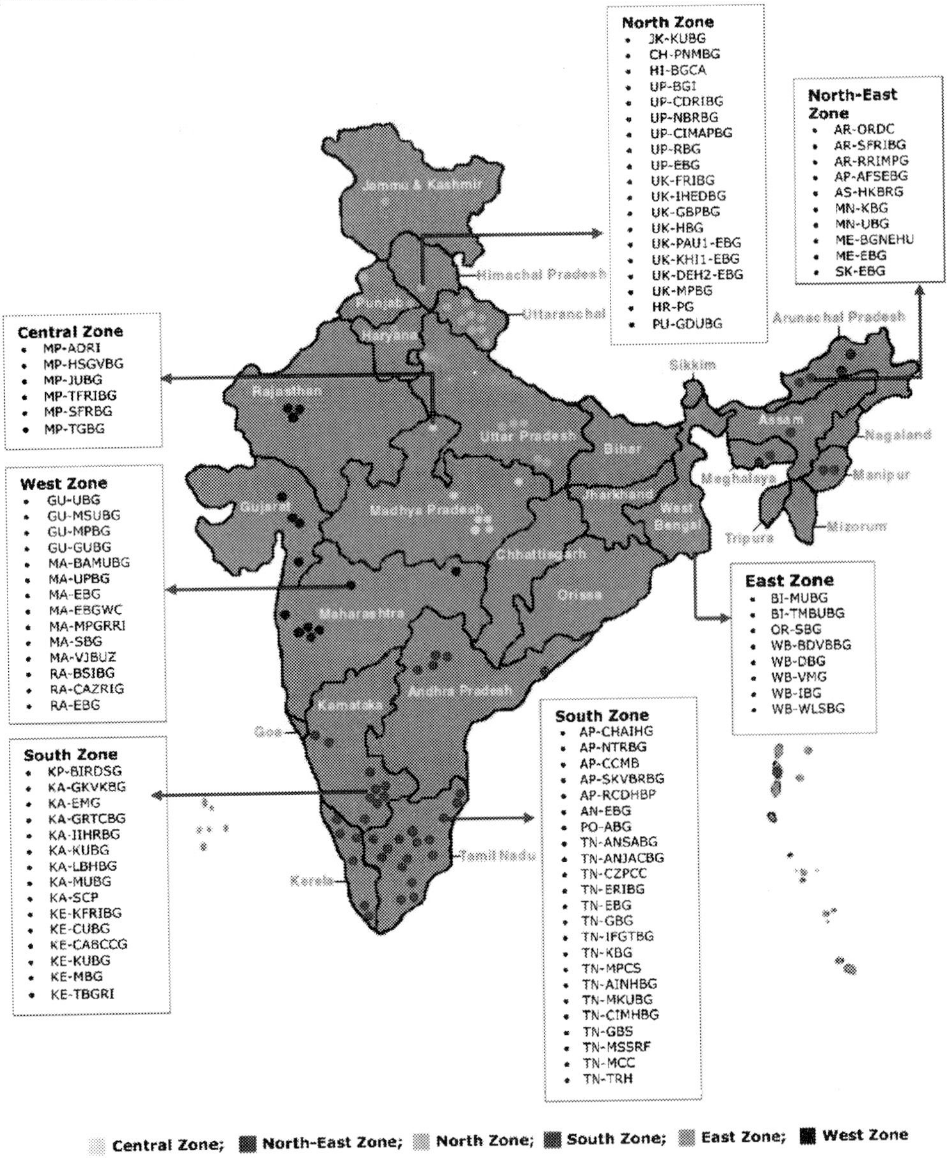

Fig. 1. Map of India showing 95 gardens registered in Indian Botanic Garden from various zones so far.

Botanic garden wise occurrence of RET species is presented in Table 4 and the study reveals maximum 68 species (12.04%) in Coimbatore Zoological Park

& Conservation Centre, Coimbatore followed by 67 species (11.86%) in Experimental Botanic Garden Eastern Circle (Shillong), Barapani, 39 species (6.90%) in Tipi, Orchid Research Centre, West Kameng, 38 species (6.73%) in Malabar Botanical Garden, Kozhikode, 34 species (6.02%) in Experimental Botanic Garden Western Circle, Mundhwa, Pune, 33 species (5.84%) in Experimental Botanic Garden Southern Circle, Yercaud, Salem, 26 species (4.60%) in Botanical Garden "Dr. H.S. Gour Vishwavidyalaya", Sagar, 25 species (4.42%) in Calcutta Wildlife Society, Kolkata, 21 species (3.72%) in SFRI Arboretum, State Forest Research Institute, Jabalpur, 21 species (3.72%) in Tropical Botanic Garden and Research Institute, Thiruvananthapuram and minimum 1 species (0.18%) in Experimental Botanic Garden Arid Zone Circle, Jodhpur. An analysis of habit of all 565 species is presented in Table 5. It shows maximum 47.08% of herbs followed by 33.63% in trees, 13.81% in shrubs, 2.12% in climber, 1.42% in herbaceous climber, 0.53% in herbaceous creeper, 0.53% in creeper, 0.35% shrub climber, 0.35% rhizome and minimum 0.18% in tuber in various botanic gardens of India.

Table 3. Showing list of rare endangered and threatened (RET) species, habits & botanic garden code etc. of India's botanic gardens

Group/Family	Plant species	Habit code	Botanic garden code
PTERIDOPHYTES			
BLECHNACEAE	*Brainea insignis* (Hook.) J.Sm.	H	ME-SHI1-EBG
CYATHEACEAE	*Alsophila crinita* Hook.	H	TN-TIR1-TRH
	Cyathea crinita Copel.	T	TN-CHE2-GBS
	C. nilgirensis Holttum	T	TN-COI1-CZPCC
DIPTERIDACEAE	*Dipteris wallichii* (R.Br.) T.Moore	H	AP-ITA-AFSEBG
DRYOPTERID-ACEAE	*Dryopteris oppositifolia* (Hook.) C.Chr.	H	TN-CHE2-GBS
EQUISETACEAE	*Equisetum arvense* L.	H	TN-SAL1-EBG
LYCOPODIACEAE	*Lycopodium japonicum* Thunb.	H	ME-SHI1-EBG
	L. squarrosum G.Forst.	H	ME-SHI1-EBG
MARATTIACEAE	*Angiopteris evecta* (G.Forst.) Hoffm.	H	TN-SAL1-EBG
POLYPODIACEAE	*Platycerium alcicorne* Desv.	H	UK-PAU1-EBG
PSILOTACEAE	*Psilotum nudum* (L.) P.Beauv.	H	UK-PAU1-EBG
GYMNOSPERMS			
CUPRESSACEAE	*Juniperus communis* L.	S	HR-GUR1-PG
	J. macropoda Boiss.	T	HR-GUR1-PG
CYCADACEAE	*Cycas beddomei* Hort. ex Hook.f.	T	TN-SAL1-EBG
	C. circinalis L.	T	ME-SHI1-EBG
	C. pectinata Buch.-Ham.	T	AP-ITA-AFSEBG

Group/Family	Plant species	Habit code	Botanic garden code
	C. revoluta Thunb.	T	UK-PAU1-EBG
	C. rumphii Miq.	T	UK-PAU1-EBG
GINKGOACEAE	*Ginkgo biloba* L.	T	TN-SAL1-EBG
GNETACEAE	*Gnetum gnemon* L.	S	ME-SHI1-EBG
	G. ula Brongn.	T	TN-SAL1-EBG
ANGIOSPERMS			
ACANTHACEAE	*Acanthus volubilis* Wall.	H	WB-KOL3-WLSBG
	Adhatoda vasica Nees	S	MP-JAB2-SFR-BG
	A. beddomei C.B.Clarke	S	UP-LKO1-NBRIBG
	Andrographis paniculata Nees	H	MP-SAG1-HSGVBG
	Avicennia alba Blume	S	WB-KOL3-WLSBG
	A. marina (Forssk.) Vierh.	T	WB-KOL3-WLSBG
	A. officinalis L.	T	WB-KOL3-WLSBG
	Justicia assamica C.B.Clarke	H	RA-JOD3-EBG
	Nilgirianthus ciliatus (Nees) Bremek.	S	TN-COI1-CZPCC
	N. foliosus (Wight) Bremek.	S	TN-COI1-CZPCC
	N. neilgherrensis (Bedd.) Bremek.	S	TN-COI1-CZPCC
	N. wightianus (Nees) Bremek.	S	TN-COI1-CZPCC
	Strobilanthes kunthianus T.Anders.ex Benth.	S	TN-COI1-CZPCC
	Thunbergia alata Bojer ex Sims	S	MP-GWA1-JUBG
	T. mysorensis T.Anders. ex Bedd.	S	TN-COI1-CZPCC
ACORACEAE	*Acorus calamus* L.	H	MP-SAG1-HSGVBG
ALLIACEAE	*Allium blandum* Wall.	H	TN-SAL1-EBG
	A. carolinianum Delar.	H	HR-GUR1-PG
AMARANTH-ACEAE	*Aerva wightii* Hook.f.	H	TN-COI3-CIMHBG
AMARYLLID-ACEAE	*Crinum defixum* [Ker-Gawl.]	H	WB-KOL3-WLSBG
HYPOXIDACEAE	*Curculigo orchioides* Gaertn.	H	HI-SOL1-BGCA
ANACARDIACEAE	*Gluta travancorica* Bedd.	T	TN-COI1-CZPCC
	Mangifera andamanica King	T	TN-SAL1-EBG
	M. indica Blume	T	TN-CHE3-MSSRF
	Nothopegia beddomei Gamble	T	TN-SAL1-EBG
	Rhus parviflora Roxb.	S	MP-JAB2-SFR-BG
	Semecarpus anacardium Blanco	T	PO-AUR1-ABG
	S. auriculata Bedd.	T	TN-COI1-CZPCC
	S. travancorica Bedd.	T	TN-COI1-CZPCC

Group/Family	Plant species	Habit code	Botanic garden code
ANCISTROCLA-DACEAE	*Ancistrocladus heyneanus* Wall.	SCl.	TN-COI1-CZPCC
ANNONACEAE	*Polyalthia rufescens* Hook.f. & Thomson	T	TN-COI1-CZPCC
	Xylopia parviflora Benth.	T	TN-COI1-CZPCC
APIACEAE	*Angelica glauca* Edgew.	H	HI-SOL1-BGCA
	Bunium persicum B.Fedtsch.	H	HR-GUR1-PG
	Carum carvi L.	H	HR-GUR1-PG
	Centella asiatica (L.) Urb.	H	MP-GWA1-JUBG
	Pimpinella pulneyensis Gamble	H	TN-TIR1-TRH
APOCYNACEAE	*Carissa opaca* Stapf ex Haines	S	MP-JAB2-SFR-BG
	Cerbera odollam Gaertn.	T	WB-KOL3-WLSBG
	Decalepis hamiltonii Wight & Arn.	S	TN-COI1-CZPCC
	Rauvolfia serpentina Benth. ex Kurz	S	MP-SAG1-HSGVBG
	Wrightia tinctoria R.Br.	T	MP-GWA1-JUBG
APONOGETON-ACEAE	*Aponogeton appendiculatus* H.W.E.van Bruggen	H	KE-KOZ1-MBG
	A. bruggenii S.R.Yadav & R.S.Govekar	H	KE-KOZ1-MBG
	A. satarensis R.Sundararaghavan , A.R.Kulkarni & S.R.Yadav	H	KE-KOZ1-MBG
AQUIFOLIACEAE	*Ilex wightiana* Wall.	T	TN-COI1-CZPCC
ARACEAE	*Cryptocoryne ciliata* Fisch. ex Schott	H	WB-KOL3-WLSBG
	C. cognata Schott	H	KE-KOZ1-MBG
	C. consobrina Schott	H	KE-KOZ1-MBG
	C. spiralis Fisch. ex Wydler	H	KE-KOZ1-MBG
	Lagenandra meeboldii (Engl.) C.E.C.Fisch.	HCr.	KE-KOZ1-MBG
	L. nairii K.Ramamurthy & R.Rajan	RH.	KE-KOZ1-MBG
	L. toxicaria Dalzell	H	KE-KOZ1-MBG
ARALIACEAE	*Schefflera rostrata* Harms	T	TN-COI1-CZPCC
	S. wallichiana (Wight & Arn.) Harms	T	ME-SHI1-EBG
ARECACEAE	*Areca nagensis* Griff.	T	ME-SHI1-EBG
	A. triandra Roxb. ex Buch.-Ham.	T	AN-DHK-EBG
	Arenga wightii Griff	T	TN-COI1-CZPCC
	Calamus andamanicus Kurz	T	TN-SAL1-EBG
	C. thwaitesii Becc.	T	TN-CHE3-MSSRF
	Hyphaene dichotoma (Wight ex Graham) Furtado	T	UP-LKO1-NBRIBG
	Korthalsia rogersii Becc.	T	AN-DHK-EBG
	Livistona jenkinsiana Griff.	T	AP-ITA-AFSEBG

Group/Family	Plant species	Habit code	Botanic garden code
	Nypa fruticans Wurmb	T	WB-KOL3-WLSBG
	Pinanga dicksonii Blume	T	TN-COI1-CZPCC
	Trachycarpus takil Becc.	T	UK-DEH1-FRIBG
	Wallichia densiflora Mart.	T	UK-PAU1-EBG
ARISTOLOCHI-ACEAE	*Aristolochia tagala* Cham.	Cl.	TN-COI1-CZPCC
ASCLEPIADA-CEAE	*Ceropegia andamanica* Sreek., Veenak. & Prashanth	H	AN-DHK-EBG
	C. bulbosa Roxb.	Cl.	MP-GWA1-JUBG
	C. sahyadrica Ansari & B.G.Kulk.	H	MA-PUN3-UPBG
	C. thwaitesii Hook.	H	TN-TIR1-TRH
	Frerea indica Dalzell	H	UP-LKO1-NBRIBG
	Gymnema sylvestre R.Br.	H	MP-SAG1-HSGVBG
	Hoya parasitica Wall. ex Traill	Cl.	WB-KOL3-WLSBG
	H. wightii Hook.f.	Cl.	TN-TIR1-TRH
	Sarcolobus carinatus Wall.	S	WB-KOL3-WLSBG
	S. globosus Wall.	S	WB-KOL3-WLSBG
	Sarcostemma intermedium Decne.	H	TN-SAL1-EBG
	Toxocarpus beddomei Gamble	S	TN-COI1-CZPCC
	Tylophora fasciculata Thwaites	S	MP-GWA1-JUBG
	T. indica Merr.	Cl.	MP-SAG1-HSGVBG
	Utleria salicifolia Bedd. ex Hook.f.	S	TN-COI1-CZPCC
ASTERACEAE	*Artemisia amygdalina* Decne.	S	JK-SRN1-KUBG
	Catamixis baccharoides Thomson	T	UK-PAU1-EBG
	Helichrysum cutchicum (C.B.Clarke) Rao Rolla & Deshp.	H	GU-VAD1-MSUBG
	Inula racemosa Hook.f.	H	HR-GUR1-PG
	Kleinia grandiflora (Wall. ex DC.) N.Rani	H	TN-SAL1-EBG
	Peucedanum nagpurense Prain	H	MP-SAG1-HSGVBG
	Saussurea costus (Falc.) Lipsch.	H	JK-SRN1-KUBG
	S. lappa (Decne.) Sch.Bip.	H	PU-AMR1-GDUBG
	Vernonia pulneyensis Gamble	H	TN-TIR1-TRH
	V. shevaroyensis Gamble	S	TN-COI1-CZPCC
	V. fysonii Calder.	H	TN-TIR1-TRH
BALSAMINACEAE	*Impatiens grandis* Heyne	S	TN-CHE2-GBS
	I. parasitica Bedd.	H	TN-CHE2-GBS
	I. acaulis Arn.	H	TN-SAL1-EBG

Group/Family	Plant species	Habit code	Botanic garden code
	I. raziana Bhaskar & Razi	H	JK-SRN1-KUBG
	I. talbotii Baker f.	H	TN-CHE2-GBS
BEGONIACEAE	*Begonia malabarica* Buch.-Ham.	H	TN-COI1-CZPCC
	Tecomella undulata (Sm.) Seem.	T	UP-LKO1-NBRIBG
BERBERIDACEAE	*Berberis aristata* DC.	S	HI-SOL1-BGCA
	B. khasiana Ahrendt	S	ME-SHI1-EBG
	Sinopodophyllum hexandrum (Royle) T.S.Ying	H	SK-GAN1-EBG
BIGNONIACEAE	*Dolichandrone spathacea* K.Schum.	T	WB-KOL3-WLSBG
	Oroxylum indicum (L.) Benth. ex Kurz	T	PU-AMR1-GDUBG
	Radermachera xylocarpa K.Schum.	T	MP-JAB2-SFR-BG
BIXACEAE	*Cochlospermum religiosum* (L.) Alston	T	MP-SAG1-HSGVBG
BORAGINACEAE	*Cynoglossum lanceolatum* Hochst. ex DC.	H	MP-JAB2-SFR-BG
BRASSICACEAE	*Megacarpaea bifida* Benth.	H	JK-SRN1-KUBG
BURSERACEAE	*Boswellia serrata* Roxb.	T	MP-SAG1-HSGVBG
	Canarium strictum Roxb.	T	TN-COI1-CZPCC
	Commiphora wightii (Arn.) Bhandari	S	MP-STN1-ADRI
CANNABACEAE	*Urginea indica* Kunth	H	MP-SAG1-HSGVBG
CAPPARACEAE	*Capparis fusifera* Dunn	T	TN-COI1-CZPCC
	Maerua oblongifolia A.Rich.	S	TN-CHE4-MCC
CELASTRACEAE	*Celastrus paniculatus* Willd.	S	MP-JAB2-SFR-BG
	Cassine glauca (Rottb.) Kuntze	T	MP-JAB2-SFR-BG
	Salacia chinensis L.	S	TN-COI3-CIMHBG
	S. oblonga Wall.	H	TN-COI1-CZPCC
	S. reticulata Wight	S	TN-COI1-CZPCC
CLUSIACEAE	*Calophyllum apetalum* Blanco	T	KE-THV2-TBGRIG
	C. polyanthum Wall.	T	KE-THV2-TBGRIG
	Garcinia indica Choisy	T	KE-THV2-TBGRIG
	G. spicata Hook.f.	T	PO-AUR1-ABG
	G. travancorica Bedd.	T	KE-THV2-TBGRIG
	G. gummi-gutta (L.) N.Robson var. *conicarpa* (Wight) N.P.Singh	T	KE-THV2-TBGRIG
	G. talbotii Raiz.	T	TN-COI1-CZPCC
	G. wightii T.Anders.	T	KE-THV2-TBGRIG
	Mesua assamica (King & Prain) Kosterm.	T	AP-ITA-AFSEBG
	M. ferrea L.	T	ME-SHI1-EBG
COMBRETACEAE	*Anogeissus sericea* Brandis	T	UP-LKO1-NBRIBG

Group/Family	Plant species	Habit code	Botanic garden code
	Combretum decandrum Ruiz & Pav. ex G.Don	S	MP-JAB2-SFR-BG
	Lumnitzera racemosa Willd.	S	WB-KOL3-WLSBG
	Terminalia chebula Willd. ex Flem.	T	MP-SAG1-HSGVBG
	T. procera Roxb.	T	TN-SAL1-EBG
CONVOLVULA-CEAE	*Convolvulus pluricaulis* Choisy	H	PU-AMR1-GDUBG
	Evolvulus alsinoides L.	H	MP-JAB2-SFR-BG
	Operculina petaloidea (Choisy) Ooststr.	H	MP-STN1-ADRI
	O. turpethum (L.) Silva Manso	H	TN-COI3-CIMHBG
COSTACEAE	*Costus speciosus* Sm.	H	MP-SAG1-HSGVBG
CUCURBITACEAE	*Citrullus colocynthis* (L.) Schrad.	H	MP-SAG1-HSGVBG
	Luffa echinata Roxb.	HCl.	MP-SAG1-HSGVBG
CYPERACEAE	*Cyperus elatus* Griseb.	H	KE-KOZ1-MBG
	C. platyphyllus Roem. & Schult.	H	KE-KOZ1-MBG
	Fimbristylis dauciformis Govindarajalu	H	KE-KOZ1-MBG
	F. hirsutifolia E.Govindarajalu	H	KE-KOZ1-MBG
	F. humerosa E.Govindarajalu	H	KE-KOZ1-MBG
	F. perspicua E.Govindarajalu & N.Sasidharan	H	KE-KOZ1-MBG
DAVALLIACEAE	*Humata repens* (L.f.) Diels	HCr.	TN-CHE2-GBS
DIOSCOREACEAE	*Dioscorea bulbifera* Russ. ex Wall.	HCl.	MP-SAG1-HSGVBG
	D. deltoidea Wall.	HCl.	HI-SOL1-BGCA
	D. hamiltonii Hook.f.	Cr.	TN-CHE3-MSSRF
	D. hispida Dennst.	HT	MP-GWA1-JUBG
	D. oppositifolia L.	HCl.	TN-COI1-CZPCC
	D. wightii Hook.f.	Cr.	MP-GWA1-JUBG
DIPTEROCARP-ACEAE	*Dipterocarpus indicus* Bedd.	T	TN-COI1-CZPCC
DROSERACEAE	*Drosera burmanni* Vahl	H	TN-SAL1-EBG
	D. peltata Thunb.	Cl.	TN-SAL1-EBG
EBENACEAE	*Diospyros affinis* Thwaites	T	TN-CHE4-MCC
	D. bourdillonii Brandis	T	TN-COI1-CZPCC
	D. chloroxylon Roxb.	T	TN-CHE4-MCC
	D. ebenum Hiern	T	PO-AUR1-ABG
	D. humilis F.Muell.	T	TN-COI1-CZPCC
	D. malabarica Kostel.	T	TN-COI1-CZPCC
	D. melanoxylon Blume	T	TN-CHE4-MCC
	D. montana Heyne ex A.DC.	T	TN-CHE4-MCC

Group/Family	Plant species	Habit code	Botanic garden code
	D. paniculata Dalzell	T	TN-COI1-CZPCC
ELAEOCARP-ACEAE	*Elaeocarpus blascoi* Weibel	T	TN-TIR1-TRH
	E. recurvatus Corner	T	TN-COI1-CZPCC
	E. serratus L.	T	TN-COI1-CZPCC
	E. tuberculatus Roxb.	T	TN-COI1-CZPCC
	E. venustus Bedd.	T	TN-COI1-CZPCC
ERIOCAULACEAE	*Eriocaulon ansarii* Pradeep & Sunil	H	KE-KOZ1-MBG
	E. cuspidatum Dalzell	H	KE-KOZ1-MBG
	E. ensiforme C.L.C.Fischer	H	KE-KOZ1-MBG
	E. heterolepis Steud.	H	KE-KOZ1-MBG
	E. lanceolatum Miq. ex Steud.	H	KE-KOZ1-MBG
	E. palghatense R.Ansari & N.P.Balakr.	H	KE-KOZ1-MBG
	E. rhodae Fyson	H	KE-KOZ1-MBG
	E. robustum Steud.	H	KE-KOZ1-MBG
	E. sivarajanii R.Ansari & N.P.Balakr.	H	KE-KOZ1-MBG
	E. vasudevanii R.Ansari & N.P.Balakr.	H	KE-KOZ1-MBG
ERYTHROXYL-ACEAE	*Erythroxylum monogynum* Roxb.	T	TN-CHE4-MCC
EUPHORBIACEAE	*Antidesma ghaesembilla* Gaertn.	T	TN-CHE4-MCC
	Baliospermum solanifolium (J.Burm.) C.R.Suresh	S	MP-JAB2-SFR-BG
	Dimorphocalyx glabellus Thwaites	T	PO-AUR1-ABG
	Drypetes sepiaria Pax & K.Hoffm.	H	PO-AUR1-ABG
	Euphorbia epiphylloides Kurz	S	AN-DHK-EBG
	Glochidion velutinum Wight	T	TN-SAL1-EBG
	Macaranga nicobarica N.P.Balakr. & Chakr.	T	AN-DHK-EBG
	Suregada angustifolia (Müll.Arg.) Airy Shaw	T	PO-AUR1-ABG
	Syzygium formosum (Wall.) Masam.	T	ME-SHI1-EBG
FAGACEAE	*Quercus glauca* Thunb. ex A.Murray	T	UK-ALM1-IHED-BG
	Q. leucotrichophora A.Camus ex Bahadur	T	UK-ALM1-IHED-BG
FLACOURTIA-CEAE	*Hydnocarpus alpina* Wight	T	TN-COI1-CZPCC
GENTIANACEAE	*Gentiana cachemirica* Decne.	H	JK-SRN1-KUBG
	G. kurroo Royle	H	HI-SOL1-BGCA
	Swertia chirata Buch.-Ham. ex Wall.	H	PU-AMR1-GDUBG
GESNERIACEAE	*Didymocarpus ovalifolia* Wight	S	TN-COI1-CZPCC
GRAMINEAE	*Cymbopogon khasianus* (Hack.) Stapf ex Bor	H	ME-SHI1-EBG

Group/Family	Plant species	Habit code	Botanic garden code
	Melocanna bambusoides Trin.	T	ME-SHI1-EBG
GROSSULARI-ACEAE	*Itea nutans* Royle	S	UK-PAU1-EBG
ICACINACEAE	*Sarcostigma kleinii* Wight & Arn.	S	TN-CHE3-MSSRF
	Nothapodytes nimmoniana (J.Graham) Mabb.	T	TN-COI1-CZPCC
LAMIACEAE	*Coleus forskohlii* Briq.	H	UK-ALM1-IHED-BG
	Eremostachys superba Royle ex Benth.	H	UK-DEH1-FRIBG
	Hyssopus officinalis L.	H	TN-TIR1-TRH
	Orthosiphon aristatus (Blume) Miq.	H	TN-SAL1-EBG
	Plectranthus bourneae Gamble	H	UK-ALM1-IHED-BG
	Salvia lanata Stokes	S	UK-ALM1-IHED-BG
LAURACEAE	*Actinodaphne bourneae* Gamble	T	TN-TIR1-TRH
	A. lanata Meisn.	T	TN-COI1-CZPCC
	Beilschmiedia wightii Benth. & Hook.f.	T	TN-COI1-CZPCC
	Cinnamomum heyneanum Nees	T	KE-THV2-TBGRIG
	C. macrocarpum Hook.f.	T	KE-THV2-TBGRIG
	C. nicolsonianum Manilal & M.Shylaja	T	KE-THV2-TBGRIG
	C. perrottetii Meisn.	T	KE-THV2-TBGRIG
	C. riparium Gamble	T	KE-THV2-TBGRIG
	C. sulphuratum Nees	T	KE-THV2-TBGRIG
	C. travancoricum Gamble	T	KE-THV2-TBGRIG
	C. wightii Lukman.	T	KE-THV2-TBGRIG
	Cryptocarya stocksii Meisn.	T	TN-TIR1-TRH
	Persea macrantha (Nees) Kosterm.	T	TN-COI1-CZPCC
	Litsea glutinosa (Lour.) C.B.Rob.	T	MP-SAG1-HSGVBG
LECYTHIDACEAE	*Careya arborea* Roxb.	T	GU-VAD1-MSUBG
LEGUMINOSAE	*Albizia odoratissima* (L.f.) Benth.	T	TN-CHE4-MCC
	Alysicarpus belgaumensis Wight	H	MA-PUN2-EBGWC
	Bauhinia foveolata Dalzell	T	MA-PUN2-EBGWC
	B. malabarica Roxb.	T	MA-PUN2-EBGWC
	B. vahlii Wight & Arn.	Cl.	MP-GWA1-JUBG
	Butea superba Roxb.	H	MA-PUN2-EBGWC
	Caesalpinia cucullata Roxb.	Cl.	MA-PUN2-EBGWC
	Cajanus lineatus Graham	S	MA-PUN2-EBGWC
	C. sericeus (Benth. ex Bak.) Maesen	S	MA-PUN2-EBGWC
	Canavalia cathartica Thou.	H	MA-PUN2-EBGWC

Group/Family	Plant species	Habit code	Botanic garden code
	C. ensiformis DC.	H	MA-PUN2-EBGWC
	Cassia angustifolia Vahl	S	PU-AMR1-GDUBG
	Castanopsis tribuloides A.DC.	T	UK-ALM1-IHED-BG
	Clitoria biflora Dalzell	H	MA-PUN2-EBGWC
	C. ternatea L.	HCl.	MP-JAB2-SFR-BG
	Crotalaria berteroana DC.	S	TN-COI1-CZPCC
	C. filipes Benth.	H	MA-PUN2-EBGWC
	C. hirta Willd.	H	MA-PUN2-EBGWC
	C. longipes Raf.	S	TN-COI1-CZPCC
	C. ramosissima F.Muell.	H	MA-PUN2-EBGWC
	C. vestita Baker	H	MA-PUN2-EBGWC
	Dalbergia horrida Graham	S	MA-PUN2-EBGWC
	Derris ovalifolia Benth.	Cl.	PO-AUR1-ABG
	Erythrina resupinata Roxb.	T	UP-LKO1-NBRIBG
	Flemingia gracilis (Mukerjee) Sanjappa	H	MA-PUN2-EBGWC
	Glycyrrhiza glabra L.	H	PU-AMR1-GDUBG
	Geissaspis tenella Benth.	H	MA-PUN2-EBGWC
	Gymnocladus assamicus Kanjilal ex P.C.Kanjilal	T	AP-ITA-AFSEBG
	Hardwickia binata Roxb.	T	MA-PUN2-EBGWC
	Hedysarum cachemirianum Benth. ex Baker	H	JK-SRN1-KUBG
	Moullava spicata (Dalzell) Nicolson	Cl.	MA-PUN2-EBGWC
	Maniltoa polyandra Harms	T	AP-ITA-AFSEBG
	Mucuna monosperma Wall.	S	MA-PUN2-EBGWC
	Ormosia travancorica Bedd.	T	TN-COI1-CZPCC
	O. robusta Baker	T	AP-ITA-AFSEBG
	Ougeinia dalbergioides Benth.	T	MP-JAB2-SFR-BG
	Pterocarpus marsupium Roxb.	T	MP-SAG1-HSGVBG
	P. dalbergioides Roxb.	T	TN-SAL1-EBG
	P. santalinus Buch.-Ham. ex Wall.	T	TN-SAL1-EBG
	Pueraria tuberosa DC.	Cl.	MP-JAB2-SFR-BG
	Rhynchosia hirta (Andrews) Meikle & Verdc.	Cl.	MA-PUN2-EBGWC
	Saraca asoca (Roxb.) De Wilde	T	MA-PUN2-EBGWC
	Smithia agharkarii Hemadri	HCl.	MA-PUN2-EBGWC
	S. bigemina Dalzell	H	MA-PUN2-EBGWC
	S. hirsuta Dalzell	H	MA-PUN2-EBGWC
	S. purpurea Hook.	H	MA-PUN2-EBGWC

Group/Family	Plant species	Habit code	Botanic garden code
	Sophora mollis Span.	S	UK-DEH1-FRIBG
	Sphenostylis bracteata (Baker f.) J.B.Gillett	H	MA-PUN2-EBGWC
	Tephrosia purpurea Pers.	S	MP-JAB2-SFR-BG
	T. coccinea Wall.	S	MA-PUN2-EBGWC
	Uraria picta Wight	H	MP-SAG1-HSGVBG
	Utricularia malabarica Janarth. & A.N.Henry	H	KE-KOZ1-MBG
	U. nayarii Janarth. & A.N.Henry	H	KE-KOZ1-MBG
	U. praeterita P.Taylor	H	KE-KOZ1-MBG
	U. reticulata Sm.	H	KE-KOZ1-MBG
	U. smithiana Wight	H	KE-KOZ1-MBG
	Vigna angularis (Willd.) Ohwi & H.Ohashi	H	MA-PUN2-EBGWC
	V. dalzelliana (Kuntze) Verdc.	HCr.	MA-PUN2-EBGWC
	V. khandalensis (Santapau) Raghavan & Wadhwa	H	MA-PUN2-EBGWC
	V. sublobata (Roxb.) G.C.Bairiganjan , P.C.Panda , B.P.Choudhury & S.N.Patnaik	H	MA-PUN2-EBGWC
	V. trilobata (L.) Verdc.	H	MA-PUN2-EBGWC
	V. aconitifolia (Jacq.) Maréchal	H	MA-PUN2-EBGWC
LILIACEAE	*Asparagus racemosus* Willd.	S	PU-AMR1-GDUBG
	Chlorophytum borivilianum Santapau & R.R.Fernandes	H	MA-PUN3-UPBG
	C. tuberosum Baker	H	MP-SAG1-HSGVBG
	Dipcadi montanum Baker	H	TN-SAL1-EBG
	Gloriosa superba L.	H	MP-SAG1-HSGVBG
	Polygonatum verticillatum All.	RH.	HI-SOL1-BGCA
LOGANIACEAE	*Strychnos minor* Benth.	SCl.	PO-AUR1-ABG
	S. potatorum L.f.	T	MP-JAB2-SFR-BG
LYTHRACEAE	*Lawsonia inermis* L.	S	PU-AMR1-GDUBG
	Rotala cookii K.T.Joseph & V.V.Sivarajan	H	KE-KOZ1-MBG
	R. fimbriata Wight	H	KE-KOZ1-MBG
	R. macrandra Koehne	H	KE-KOZ1-MBG
	R. ritchiei Koehne	H	KE-KOZ1-MBG
	R. vasudevanii K.T.Joseph & V.V.Sivarajan	H	KE-KOZ1-MBG
	Woodfordia fruticosa Kurz	T	MP-JAB2-SFR-BG
MALVACEAE	*Grewia hirsuta* Vahl	H	MP-JAB2-SFR-BG
	Heritiera fomes Wall.	T	WB-KOL3-WLSBG
	Marattia fraxinea Sm.	H	TN-CHE2-GBS

Group/Family	Plant species	Habit code	Botanic garden code
MELASTOMA-TACEAE	*Sonerila* pulneyensis Gamble	H	TN-COI1-CZPCC
	Memecylon flavescens Gamble	S	TN-COI1-CZPCC
	Osbeckia cupularis D.Don ex Wight & Arn.	S	TN-SAL1-EBG
MELIACEAE	*Aglaia elaeagnoidea* Benth.	T	PO-AUR1-ABG
	Walsura trifoliata (A. Juss.) Harms	T	TN-CHE4-MCC
	W. trifoliolata (A. Juss.) Harms	T	PO-AUR1-ABG
	W. candollei King	T	UP-LKO1-NBRIBG
	Xylocarpus granatum K.D. Koenig	T	WB-KOL3-WLSBG
	X. mekongensis Pierre	T	WB-KOL3-WLSBG
MENISPERM-ACEAE	*Stephania hernandifolia* Walp.	S	ME-SHI1-EBG
MENYANTH-ACEAE	*Nymphoides aurantiaca* Kuntze	H	KE-KOZ1-MBG
	N. krishnakesara K.T.Joseph & Sivar.	H	KE-KOZ1-MBG
	N. sivarajanii K.T.Joseph	H	KE-KOZ1-MBG
MINISPERM-ACEAE	*Coscinium fenestratum* Colebr.	T	TN-COI1-CZPCC
MORACEAE	*Artocarpus gomezianus* Wall. ex Trécul	T	TN-COI1-CZPCC
	A. hirsutus Lam.	T	KE-THV2-TBGRIG
	Ficus angladei C.E.C.Fisch.	T	TN-TIR1-TRH
	F. beddomei King	T	TN-COI1-CZPCC
MUSACEAE	*Ensete superbum* (Roxb.) Cheesman	S	TN-COI1-CZPCC
	Musa acuminata Colla	T	TN-SAL1-EBG
	M. velutina H.Wendl. & Drude	T	ME-SHI1-EBG
MYRISTICACEAE	*Knema attenuata* Warb.	T	TN-SAL1-EBG
	K. andamanica (Warb.) W.J.de Wilde	S	TN-SAL1-EBG
	Myristica fatua Houtt.	T	TN-COI1-CZPCC
	M. dactyloides Gaertn.	T	TN-COI1-CZPCC
	M. andamanica Hook.f.	T	AN-DHK-EBG
	M. malabarica Lam.	T	TN-COI1-CZPCC
MYRSINACEAE	*Aegiceras corniculatum* Blanco	S	WB-KOL3-WLSBG
	Embelia ribes Burm.f.	S	TN-CHE3-MSSRF
	E. tsjeriam-cottam A.DC.	T	MP-JAB2-SFR-BG
MYRTACEAE	*Rhodomyrtus tomentosa* Hassk.	S	TN-SAL1-EBG
	Syzygium bourdillonii (Gamble) N.C.Rathakrishnan & N.C.Nair	T	KE-THV2-TBGRIG

Group/Family	Plant species	Habit code	Botanic garden code
	S. chavaran Gamble	T	KE-THV2-TBGRIG
	S. densiflorum Wall.	T	TN-COI1-CZPCC
	S. laetum (Ham.) Gandhi	T	KE-THV2-TBGRIG
	S. mundagam (Bourd.) Chithra	T	KE-THV2-TBGRIG
	S. parameswaranii M.Mohanan & A.N.Henry	T	KE-THV2-TBGRIG
	S. travancoricum Gamble	T	TN-COI1-CZPCC
	S. zeylanicum (L.) DC.	T	KE-THV2-TBGRIG
NEPENTHACEAE	*Nepenthes khasiana* Hook.f.	S	AP-ITA-AFSEBG
NYMPHAEACEAE	*Nymphaea pygmaea* [Dryand.]	H	ME-SHI1-EBG
OLEACEAE	*Chionanthus mala-elengi* (Dennst.) P.S.Green	T	TN-CHE4-MCC
	Jasminum mesnyi Hance	S	ME-SHI1-EBG
ORCHIDACEAE	*Aenhenrya rotundifolia* (Blatt.) C.S. Kumar & F.N. Rasm.	H	AR-WEK1-ORDC
	Anoectochilus roxburghii Lindl.	H	ME-SHI1-EBG
	Ascocentrum miniatum Schltr.	H	AR-WEK1-ORDC
	Bulbophyllum hirtum Lindl.	H	ME-SHI1-EBG
	B. pectinatum Finet	H	AR-WEK1-ORDC
	Brachycorythis galeandra (Rchb.f.) Summerh.	T	ME-SHI1-EBG
	Calanthe alismaefolia Lindl.	H	AR-WEK1-ORDC
	C. anthropophora Ridl.	H	AR-WEK1-ORDC
	Coelogyne cristata Lindl.	H	ME-SHI1-EBG
	C. viscosa Rchb.f.	H	ME-SHI1-EBG
	C. fuscescens Lindl.	H	ME-SHI1-EBG
	Cymbidium cochleare Lindl.	H	UK-PAU1-EBG
	C. devonianum Paxton	H	ME-SHI1-EBG
	C. eburneum Lindl.	H	AR-WEK1-ORDC
	C. hookerianum Rchb.f.	H	UK-PAU1-EBG
	C. iridioides D.Don	H	TN-SAL1-EBG
	C. macrorhizon Lindl.	H	ME-SHI1-EBG
	C. lowianum Rchb.f.	H	AR-WEK1-ORDC
	C. parishii Rchb.f.	H	AR-WEK1-ORDC
	C. tigrinum Parish ex Hook.	H	AR-WEK1-ORDC
	Dendrobium aqueum Lindl.	H	AR-WEK1-ORDC
	D. crepidatum Lindl. & Paxton	H	AR-WEK1-ORDC
	D. draconis Rchb.f.	H	AR-WEK1-ORDC
	D. heterocarpum Wall. ex Lindl.	H	AR-WEK1-ORDC

Group/Family	Plant species	Habit code	Botanic garden code
	D. parishii Rchb.f.	H	AR-WEK1-ORDC
	D. seidenfadenii Senghas & L.Bockemuhl	H	AR-WEK1-ORDC
	D. thyrsiflorum Rchb.f. ex André	H	AR-WEK1-ORDC
	D. wattii Rchb.f.	H	AR-WEK1-ORDC
	D. williamsonii Day & Rchb.f.	H	AR-WEK1-ORDC
	D. nobile Lindl.	H	UP-LKO1-NBRIBG
	D. cathcartii Hook.f.	H	ME-SHI1-EBG
	D. formosum Roxb.	H	ME-SHI1-EBG
	D. grande Hook.f.	H	AN-DHK-EBG
	D. infundibulum Lindl.	H	ME-SHI1-EBG
	D. normale Falc.	H	UK-PAU1-EBG
	D. plicatile Lindl.	H	AN-DHK-EBG
	D. stuposum Lindl.	H	ME-SHI1-EBG
	Diplomeris pulchel D.Don	H	AR-WEK1-ORDC
	Disperis zeynica Trim.	H	TN-SAL1-EBG
	Doritis pulcherrima Lindl.	H	AR-WEK1-ORDC
	Dactylorhiza hatagirea (D.Don) Soó	H	AR-WEK1-ORDC
	Eria apertiflora Summerh.	H	AR-WEK1-ORDC
	Eulophia epidendraea C.E.C.Fisch.	H	PO-AUR1-ABG
	E. herbacea Lindl.	H	MP-SAG1-HSGVBG
	Epigeneium amplum (Lindl. ex Wall.) Summerh.	H	ME-SHI1-EBG
	E. fuscescens (Griff.) Summerh.	H	ME-SHI1-EBG
	Eria acervata Lindl.	H	ME-SHI1-EBG
	E. biflora Lindl.	H	ME-SHI1-EBG
	E. muscicola Lindl.	H	ME-SHI1-EBG
	E. paniculata Lindl.	H	ME-SHI1-EBG
	E. stricta Lindl.	H	ME-SHI1-EBG
	Esmeralda clarkei Rchb.f.	H	AR-WEK1-ORDC
	Flickingeria macraei (Lindl.) Bennet	H	ME-SHI1-EBG
	Goodyera foliosa Benth. ex Hook.f.	H	ME-SHI1-EBG
	Habenaria khasiana Hook.f.	H	AR-WEK1-ORDC
	H. pectinata D.Don	H	AR-WEK1-ORDC
	H. reniformis Hook.f.	H	AR-WEK1-ORDC
	Holcoglossum amesianum (Rchb.f.) Christenson	H	AR-WEK1-ORDC
	Hetaeria gardneri Benth. & Hook.f.	H	TN-CHE3-MSSRF
	Hygrochilus parishii Pfitzer	H	AR-WEK1-ORDC

Group/Family	Plant species	Habit code	Botanic garden code
	Ipsea malabarica Hook.f.	H	AN-DHK-EBG
	Katherinea navicularis Balakr. & Sud.Chowdhury	H	ME-SHI1-EBG
	Liparis plantaginea Lindl.	H	ME-SHI1-EBG
	Malaxis acuminata D.Don	H	ME-SHI1-EBG
	Monomeria barbata Lindl.	H	AR-WEK1-ORDC
	Oberonia bicornis Lindl.	T	AR-WEK1-ORDC
	O. rufilabris Lindl.	H	ME-SHI1-EBG
	Panisea tricallosa Rolfe	H	ME-SHI1-EBG
	Paphiopedilum charlesworthii Pfitzer	H	AR-WEK1-ORDC
	P. hirsutissimum (Lindl. ex Hook.) Stein.	H	AR-WEK1-ORDC
	P. insigne Pfitzer	H	AR-WEK1-ORDC
	P. spicerianum Pfltz.	H	UK-PAU1-EBG
	P. venustum Pfitzer	H	ME-SHI1-EBG
	P. villosum (Lindl.) Stein	H	ME-SHI1-EBG
	P. wardii Summerh.	H	AR-WEK1-ORDC
	Phaius mishmensis Rchb.f.	H	ME-SHI1-EBG
	Phalaenopsis speciosa Rchb.f.	H	AN-DHK-EBG
	Pholidota wattii King & Pantl.	H	ME-SHI1-EBG
	Pleione hookeriana Kuntze	H	UK-PAU1-EBG
	Rhynchostylis retusa Blume	H	ME-SHI1-EBG
	Renanthera imschootiana Rolfe	H	ME-SHI1-EBG
	Tainia latifolia Benth. ex Hook.f.	H	ME-SHI1-EBG
	T. minor Hook.f.	H	ME-SHI1-EBG
	Thelasis bifolia Hook.f.	H	AR-WEK1-ORDC
	T. khasiana Hook.f.	H	AR-WEK1-ORDC
	Thunia alba Rchb.f.	H	ME-SHI1-EBG
	Trichoglottis tenera Rchb.f.	H	TN-TIR1-TRH
	Vanda bicolor Griff.	H	ME-SHI1-EBG
	V. coerulea Griff. ex Lindl.	H	TN-SAL1-EBG
	V. jainii A.S.Chauhan	H	AR-WEK1-ORDC
	V. pumila Hook.f.	H	AR-WEK1-ORDC
	V. spathulata Spreng.	H	PO-AUR1-ABG
	V. stangeana Rchb.f.	H	SK-GAN1-EBG
	V. testacea Rchb.f.	H	ME-SHI1-EBG
	Vanilla pilifera Holttum	H	AR-WEK1-ORDC
	V. planifolia Andrews	H	UP-LKO1-NBRIBG

Group/Family	Plant species	Habit code	Botanic garden code
PANDANACEAE	*Pandanus andamanensis* Hort. ex Balf.f.	T	AN-DHK-EBG
PAPAVERACEAE	*Meconopsis latifolia* Prain	H	JK-SRN1-KUBG
PIPERACEAE	*Piper longum* L.	HCl.	AN-DHK-EBG
PITTOSPOR-ACEAE	*Pittosporum eriocarpum* Royle	S	UK-PAU1-EBG
PLANTAGIN-ACEAE	*Plantago ovata* Phil.	H	PU-AMR1-GDUBG
PLUMBAGIN-ACEAE	*Plumbago zeylanica* L.	S	MP-SAG1-HSGVBG
POACEAE	*Bambusa balcooa* Roxb.	T	ME-SHI1-EBG
	B. glaucescens (Willd.) Merr.	T	ME-SHI1-EBG
	B. longispiculata Gamble	T	ME-SHI1-EBG
	B. pallida Munro	T	AP-ITA-AFSEBG
	B. tulda Benth.	T	AP-ITA-AFSEBG
	Bentinckia condapanna Berry ex Roxb.	T	TN-SAL1-EBG
	B. nicobarica Hort.Sand.	T	TN-SAL1-EBG
	Ochlandra beddomei Gamble	T	TN-COI1-CZPCC
	O. travancorica Gamble	T	TN-COI1-CZPCC
POLYGONACEAE	*Rheum australe* D.Don	H	HI-SOL1-BGCA
RANUNCUL-ACEAE	*Aconitum atrox* Walp.	H	HR-GUR1-PG
	A. ferox Wall.	H	SK-GAN1-EBG
	A. heterophyllum Wall.	H	HR-GUR1-PG
	A. violaceum Jacquem. ex Stapf	S	JK-SRN1-KUBG
	Aquilegia nivalis Falc. ex Baker	S	JK-SRN1-KUBG
	Coptis teeta Wall.	S	ME-SHI1-EBG
	Thalictrum foliolosum Hook.f. & Thomson	H	MP-SAG1-HSGVBG
RHAMNACEAE	*Ziziphus oenoplia* (L.) Mill.	S	MP-JAB2-SFR-BG
RHIZOPHOR-ACEAE	*Bruguiera parviflora* (Roxb.) Griff.	T	WB-KOL3-WLSBG
	B. sexangula Poir.	T	WB-KOL3-WLSBG
	Kandelia candel Druce	T	WB-KOL3-WLSBG
RUBIACEAE	*Gardenia gummifera* L.f.	T	TN-COI1-CZPCC
	G. latifolia [Soland.]	T	TN-TIR1-TRH
	Mitragyna parvifolia Korth.	T	MP-SAG1-HSGVBG
	Neonauclea gageana Merr.	H	AN-DHK-EBG
	Pavetta wightii Hook.f.	T	TN-COI1-CZPCC

Group/Family	Plant species	Habit code	Botanic garden code
	Psydrax ficiformis (Hook.f.) Bridson	T	TN-CHE4-MCC
	Rubia cordifolia Thunb.	HCl.	MP-SAG1-HSGVBG
	Tricalysia sphaerocarpa Gamble	T	PO-AUR1-ABG
RUTACEAE	*Chloroxylon swietenia* DC.	T	PO-AUR1-ABG
	Hedyotis beddomei Hook.f.	S	TN-COI1-CZPCC
	Pamburus missionis Swingle	T	PO-AUR1-ABG
	Skimmia laureola (DC.) Dcne.	S	HI-SOL1-BGCA
	Vepris bilocularis Engl.	T	TN-COI1-CZPCC
SALVADORACEAE	*Salvadora oleoides* Decne.	T	MP-STN1-ADRI
	S. persica L.	T	WB-KOL3-WLSBG
SANTALACEAE	*Santalum album* L.	T	PU-AMR1-GDUBG
SAPINDACEAE	*Acer laevigatum* Wall.	T	ME-SHI1-EBG
	Aesculus assamica Griff.	T	ME-SHI1-EBG
	Lepisanthes tetraphylla Radlk.	S	PO-AUR1-ABG
SAPOTACEAE	*Manilkara hexandra* Dubard	T	TN-CHE4-MCC
SAXIFRAGACEAE	*Bergenia ciliata* A.Braun ex Engl.	H	SK-GAN1-EBG
	B. ligulata (Wall.) Engl.	H	UK-ALM1-IHED-BG
SCHIZAEACEAE	*Anemia schimperiana* C.Presl	H	TN-SAL1-EBG
SCROPHULARI-ACEAE	*Bacopa monnieri* (L.) Pennell	H	PU-AMR1-GDUBG
	Lagotis cashmeriana Rupr.	H	JK-SRN1-KUBG
	Picrorhiza kurrooa Royle	H	SK-GAN1-EBG
SMILACACEAE	*Smilax wightii* A.DC.	S	TN-COI1-CZPCC
SOLANACEAE	*Solanum indicum* L.	S	HR-GUR1-PG
	S. khasianum C.B.Clarke	H	ME-SHI1-EBG
	S. trilobatum Noronha	Cr.	WB-KOL3-WLSBG
	Hyoscyamus niger L.	H	TN-COI3-CIMHBG
	Withania somnifera Dunal	S	PU-AMR1-GDUBG
SONNERATIA-CEAE	*Sonneratia apetala* Buch.-Ham.	T	WB-KOL3-WLSBG
	S. caseolaris Druce	T	WB-KOL3-WLSBG
	S. griffithii Kurz	T	WB-KOL3-WLSBG
STERCULIACEAE	*Helicteres isora* L.	S	MP-GWA1-JUBG
	Pterospermum reticulatum Wight & Arn.	T	TN-COI1-CZPCC
	P. xylocarpum (Gaertn.) Santapau & Wagh	T	PO-AUR1-ABG
	Sterculia urens Roxb.	T	MP-SAG1-HSGVBG
SYMPLOCACEAE	*Symplocos microphylla* Wight	S	TN-COI1-CZPCC

Group/Family	Plant species	Habit code	Botanic garden code
TAXACEAE	*Taxus baccata* Hook.	T	TN-SAL1-EBG
	T. wallichiana Zucc.	T	HI-SOL1-BGCA
THEACEAE	*Schima wallichii* Choisy	T	ME-SHI1-EBG
THELYPTERI-DACEAE	*Pronephrium triphyllum* (Sw.) Holttum	H	TN-CHE2-GBS
THYMELAE-ACEAE	*Aquilaria malaccensis* Lam.	T	AP-ITA-AFSEBG
ULMACEAE	*Trema orientalis (*L.) Blume	T	MP-JAB2-SFR-BG
UMBELLIFERAE	*Heracleum candicans* Wall. ex DC.	H	HI-SOL1-BGCA
VALERIAN-ACEAE	*Nardostachys jatamansi* DC.	H	PU-AMR1-GDUBG
	Valeriana wallichii DC.	H	UK-ALM1-IHED-BG
	V. jatamansi Wall.	H	SK-GAN1-EBG
VERBENACEAE	*Clerodendrum serratum* Moon	S	MP-STN1-ADRI
VITACEAE	*Cayratia pedata* Gagnep.	S	TN-COI1-CZPCC
ZINGIBERACEAE	*Amomum hypoleucum* Thwaites	H	TN-SAL1-EBG
	A. microstephanum Baker	H	KE-MAL1-CUBG-BG
	Boesenbergia siphonantha (King ex Baker) M.Sabu	H	AN-DHK-EBG
	Curcuma caesia Roxb.	H	MP-GWA1-JUBG
	C. vamana Sabu M. & J.K.Mangaly	H	KE-MAL1-CUBG-BG
	C. zedoaria Rosc.	H	MP-SAG1-HSGVBG
	Hedychium dekianum A.S.Rao & Verma	H	ME-SHI1-EBG
	H. elwesii Baker	H	ME-SHI1-EBG
	H. gardnerianum Wall. ex Spreng.	H	ME-SHI1-EBG
	H. griffithianum Wall.	H	ME-SHI1-EBG
	H. hookeri C.B.Clarke ex Baker	H	ME-SHI1-EBG
	H. rubrum A.S.Rao & Verma	H	KE-MAL1-CUBG-BG
	H. greenii W.W.Sm.	H	KE-MAL1-CUBG-BG
	Kaempferia evansii Blatt.	H	ME-SHI1-EBG
	K. scaposa Benth. & Hook.f.	H	KE-MAL1-CUBG-BG
	Paracautleya bhatii R.M.Sm.	H	KE-MAL1-CUBG-BG
	Zingiber wightianum Thwaites	H	KE-MAL1-CUBG-BG
ZYGOPHYLLA-CEAE	*Balanites roxburghii* Planch.	S	MP-JAB2-SFR-BG
	Tribulus rajasthanensis Bhandari & V.S.Sharma	H	GU-VAD1-MSUBG

Note: H= Herb; S= Shrub; T= Tree; Cl.= Climber; Cr.= Creeper; HCl.= Herbaceous Climber; HCr.= Herbaceous Creeper; HT= Herb Tuber; RH= Rhizome; SCl.= Shrub climber

Table 4. Showing total number of Rare Endangered and Threatened species in botanic gardens of India

S.No.	Botanic garden codes	Zone	Total species	%
1.	TN-COI1-CZPCC	SZ	68	12.04
2.	ME-SHI1-EBG	NEZ	67	11.86
3.	AR-WEK1-ORDC	NEZ	39	6.90
4.	KE-KOZ1-MBG	SZ	38	6.73
5.	MA-PUN2-EBGWC	WZ	34	6.02
6.	TN-SAL1-EBG	SZ	33	5.84
7.	MP-SAG1-HSGVBG	CZ	26	4.60
8.	WB-KOL3-WLSBG	EZ	25	4.42
9.	KE-THV2-TBGRIG	SZ	21	3.72
10.	MP-JAB2-SFR-BG	CZ	21	3.72
11.	PO-AUR1-ABG	SZ	17	3.01
12.	AN-DHK-EBG	SZ	14	2.48
13.	PU-AMR1-GDUBG	NZ	13	2.30
14.	TN-TIR1-TRH	SZ	13	2.30
15.	UK-PAU1-EBG	NZ	13	2.30
16.	TN-CHE4-MCC	SZ	12	2.12
17.	AP-ITA-AFSEBG	SZ	11	1.95
18.	HI-SOL1-BGCA	NZ	10	1.77
19.	JK-SRN1-KUBG	NZ	10	1.77
20.	MP-GWA1-JUBG	CZ	10	1.77
21.	HR-GUR1-PG	NZ	9	1.59
22.	UP-LKO1-NBRIBG	NZ	9	1.59
23.	TN-CHE2-GBS	SZ	8	1.42
24.	UK-ALM1-IHED-BG	NZ	8	1.42
25.	KE-MAL1-CUBG-BG	SZ	7	1.24
26.	SK-GAN1-EBG	NEZ	6	1.06
27.	TN-CHE3-MSSRF	SZ	6	1.06
28.	MP-STN1-ADRI	CZ	4	0.71
29.	TN-COI3-CIMHBG	SZ	4	0.71
30.	GU-VAD1-MSUBG	WZ	3	0.53
31.	UK-DEH1-FRIBG	NZ	3	0.53
32.	MA-PUN3-UPBG	WZ	2	0.35
33.	RA-JOD3-EBG	WZ	1	0.18
	Total		565	

Note: SZ= South Zone, NZ= North Zone, WZ= West Zone, NEZ= North East Zone, EZ= East Zone, CZ= Central Zone

Table 5. Spectrum of habit of RET species in botanic gardens of India

Habit	Total no. of Species	%
Herb	266	47.08
Tree	190	33.63
Shrub	78	13.81
Climber	12	2.12
Herbaceous Climber	8	1.42
Herbaceous Creeper	3	0.53
Creeper	3	0.53
Shrub climber	2	0.35
Rhizome	2	0.35
Tuber	1	0.18
Total	**565**	

Botanic garden zone wise occurrence of RET species is presented in Figure 2 and study reveals maximum 45% RET species in South Zone followed by 20% in North East Zone, 13% in North Zone, 11% in Central Zone, 7% in West Zone and minimum 4% RET species in East Zone.

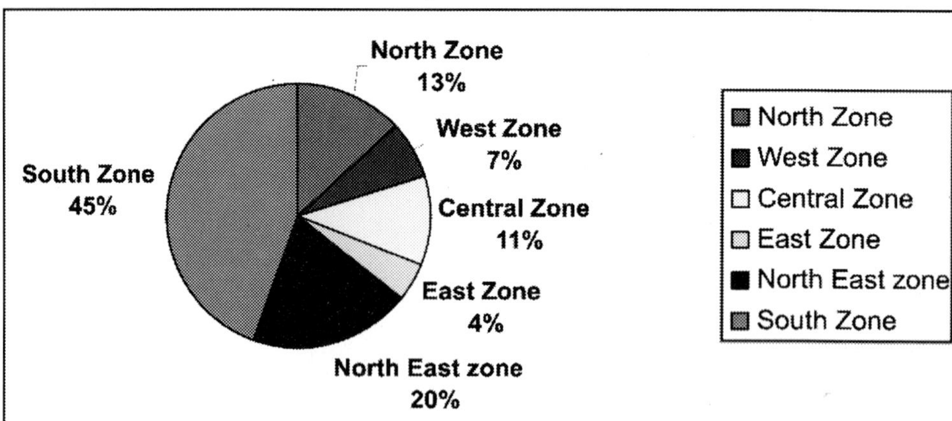

Fig. 2. Showing occurrence of Rare, Endangered and Threatened species in six zones viz., North Zone, West Zone, Central Zone, East Zone, North East Zone and South Zone.

The work of the IBGN gives special attention and concern to conservation of indigenous plant diversity. It will:

- facilitate the preparation of an Action Plan for Botanic Gardens in India, to guide the work of the Network and to assist in highlighting priority tasks for individual institutions and for the Network itself. This Action Plan will include measurable targets to provide a focus for the work and achievements of botanic gardens over the coming years.
- develop the means to ensure that data and information in support of the implementation of the Action Plan should be widely accessible in

electronic and other published forms (website, database, newsletters, technical manuals and other publications) while at the same time safeguarding the intellectual property rights of data providers.

- be assisted by the establishment of a national programme on the conservation of indigenous plants using all appropriate methodologies including the integration of *in situ* and *ex situ* techniques and the recovery of threatened plants.
- play an active role in promoting and representing the interests of botanic gardens in India to the broader environmental community. It should also support the commitment of botanic gardens in India to supporting sustainable livelihoods in their local regions by implementing appropriate community-based programmes.

The establishment of the Indian Botanic Gardens Network is first of its kind for India, its website, databases, and online facilities have already provided Indian botanic gardens with various opportunities including: I-effectively registering their existence and resources; II-sharing the information of others, and; III-receiving world-wide exposure of their organizations via the internet. Besides this, the codes for each garden will act as a unique identifier for each botanical garden and these will help in the digitization of collections available in India's botanic gardens. IBGN has a unique feature of individual membership in the network, which will make it more versatile in future. As a better understanding of the botanic garden community in India emerges, it will be possible to take good advantage of them as both a conservation, and environmental education resource. In addition, the development of a national Action Plan for Botanic Gardens will greatly assist them to operate in a way they would be unable to do as individual botanic gardens. Overall, this will greatly help India to implement the targets of the Global Strategy for Plant Conservation and India's commitment in other treatise and conventions on biodiversity.

After the formation of Indian Botanic Garden Network it is essential to strengthen it further so that it can serve as an important resource for meeting the challenges posed by climate change. IBGN has been developed using a common format sharing of resources among Indian botanic gardens and it will be expanded further on the following lines so as to tackle climate change:

(i) Value addition to data of RET species with respect to nomenclature, description, uses, images and genomic information, etc.

(ii) Further additions of data on species other than RET growing in botanic gardens and thus make comprehensive account of species diversity in Indian botanic gardens.

(iii) Data on assessment of propagation success in each botanic garden and emphasis for developing the propagation protocol.

(iv) Data on flowering time of various species.

IBGN will serve as an effective important resource in tackling climate change and this can be achieved after setting up following goals:

i. Strengthening the network by adding other botanical gardens, repositories: It is essential to strengthen the network by adding other botanical gardens and more information for educating the public and thus creating awareness and work is to be done for safety of India's plant species through *ex-situ* conservation. A nationally coordinated network of botanic gardens and incorporating further work in botanic gardens, is a critical component of a more efficient safety network for Indian plant species and an essential part of an integrated national response to the impacts of climate change. It will provide an opportunity to jointly prioritize conservation objectives, develop an integrated national infrastructure, develop standards for representative genetic diversity collections etc.

ii. Deliver comprehensive and integrated and easily accessible information about India's plant species growing in botanic gardens: The combined knowledge and expertise held by botanic gardens, in associationship with herbaria, will make a substantial contribution to the understanding of plant adaptation to climate change and support plant conservation in the wild and in botanic gardens. The botanical expertise in plant diversity and taxonomy held in botanic gardens and herbaria underpins the current state of knowledge about plant diversity. A few botanic gardens and herbaria also have considerable research expertise and facilities for molecular analysis for population and evolutionary studies as well as seed biology, plant researchers and other research strengths.

iii. Establish a monitoring programme of plant responses to environment changes: India's botanic gardens are the source of plant flowering times, which is a key indicators of response to environmental changes. In addition, botanic gardens have significant expertise and knowledge of plant flowering and seasonal studies. They are well placed to develop methodologies for monitoring the impact of climate change on plants for botanic gardens and conservation agencies in India.

iv. Create national awareness of climate change and facilitate effective responses: Botanic Gardens have already established effective patterns of community engagement and already conduct a broad range of education, botanical information, and interpretation and community awareness activities. Development of a national climate change community awareness agenda to focus collaboration will increase the power and impact of current programs. National coordination will also facilities sharing of resources and expertise between capital cities and regional gardens. Consistent programs and messages that are national in scope can be more effectively applied at major and regional botanic gardens.

It will not be out of place to mention that the present study is first of its kind effort to compile information on botanic gardens and all RET species growing in botanic gardens of India and put these in one place and made accessible through Indian Botanic Garden Network on the basis of inputs provided by respective gardens. It will be regularly updated with the inputs received from remaining botanic gardens and experts and which may lead to increase or decrease in this total RET species because a few botanic gardens do not have taxonomic expertise and hence possibility of confusions in identification of species can not be ruled out[10].

Acknowledgements

Authors are thankful to Dr. C.S. Nautiyal, Director, NBRI, Lucknow for his help in resumption of Bioinformatics Lab, which resulted this publication.

References

1. http://www.bgci.org.uk/conservation/
2. Hurka H, Conservation genetics and the role of botanical gardens in Conservation Genetics (eds. Loeschcke, V., Tomsuk, J. and Jain, S.K.), Birkha (1994).
3. Wyse Jackson Peter S, Botanic Gardens and the Convention on Biological Diversity. Canadian Botanical Conservation Network Newsletter Special Issue, March 1997, 1(4) (http://www.rbg.ca/cbcn/en/publications/ newsletters/0104.html) (1997)..
4. Kumar S, Indian Botanic Garden Network Newsletter, BGCI, UK (2003). 1(1): pp. 1-12.
5. Kumar S, Indian Botanic Garden Network Newsletter, BGCI, UK (2003). 1(2): pp. 1-20.
6. Kumar S, Indian Botanic Garden Network Newsletter, BGCI, UK (2003). 1(3): pp. 1-20.
7. Kumar S, Indian Botanic Garden Network Newsletter, BGCI, UK (2004). 2(1): pp. 1-40.
8. Kumar S, Indian Botanic Garden Network Newsletter, BGCI, UK (2005). 3(1): pp. 1-36.
9. Anonymous, Botanic Garden-A National Facility, NBRI, Lucknow, (2007). pp. 1-16.
10. Anonymous, National Strategy and Action Plan for the Role of Australia's Botanic Gardens in Adapting to Climate Change, (2008). pp. 1-27.
11. Vernon H Heywood, The role of botanic gardens as resource and introduction centres in the face of global change. In "Special issue: EuroGard V: Botanic Gardens in the Age of Climate Change/Guest edited by Leif Schulman" Biodiversity and Conservation (2010). 20 (2): pp. 221-239.

Hypoxia in Indian estuaries — Krishna and Godavari estuarine systems — A case study

V. Rajani Kumari & I. Mrutyunjaya Rao

Department of PNCO, School of Chemistry, Andhra University,
Visakhapatnam – 530 003. India
[E-mail: imrao_ivaturi@yahoo.co.in]

Introduction

Estuaries, as a constituent of coastal margins, are important ecological systems for their unique biogeochemical processes influenced by both lithological and oceanic conditions associated with natural and anthropogenic activities. India, with a long coast line of about 7500 km has 14 major, 44 medium and 162 minor rivers discharging annually about 1675 km^3 fresh water[1] and 2.873×10^{12} kg of suspended solids[2] into the Indian ocean. The important major rivers namely, the Ganges, Mahanadi, Godavari, Krishna and Kaveri on the east coast and the Narmada and the Tapti on the west coast contribute (812 km^3) nearly half of the total fresh water discharge transporting 1.119×10^{12} kg of the silt sea[3]. Further, Indian estuaries are greatly influenced by human activities such as over exploitation of aquatic resources, discharges of domestic, agricultural and industrial wastes, establishment of irrigation or power generation projects leading to control of natural river flows, etc., resulting in changes in their water quality and health status, biodiversity and other hydrodynamic and biogeochemical characteristics.

Dissolved oxygen, a very important water quality parameter is taken as an index as it is most influenced by varying physical, topographical, sedimentological, chemical and biological conditions prevailing in the Indian estuaries to identify water quality status such as oxic, hypoxic or anoxic in nature.

Keeping these aspects in view, the distribution of dissolved oxygen in estuarine environment of Krishna estuary and its water quality status with reference to dissolved oxygen is presented in this paper considering the degree of under saturation computed from the observed and equilibrium concentration levels of dissolved oxygen.

Materials and methods

Study area and station locations

Krishna estuary with about 45 km in length including its three distributaries covers an area of about 320 km^2. It is essentially a shallow one with a mean depth of 5 to 7 m and its tidal portion extends up to 39 km upstream near Penumudi. The estuary has a well developed sandy coast that experiences long shore drift, built-up spits and barriers across river mouths creating a coastal lagoon. It is a meso tidal estuary with a tide range of 2-3 m. The dissolved and particulate loads of the Krishna river are derived from a variety of rocks, catchment areas, domestic agricultural and industrial wastes due to human activity. A good number of irrigation projects that have come up on the Krishna river have drastically restricted to the natural course of the river discharge and affect the geography of the estuarine system. Further, the estuarine system is fringed with extensive mangrove vegetation.

Godavari is the second largest river in India. It opens into the Bay of Bengal on the east coast after flowing across the peninsular India over a distance of approximately 1500 km. The annual discharge of the Godavari river into the Bay of Bengal is 1.05×10^{14} (UNESCO, 1971)[4]. The river rises (16° – 18° N, 73° – 83°3′ E) at Triambak near Nasik district about 112.7 km northeast of Bombay. Several small tributaries including the Manjira and the Kudaliar rivers arise in the southern part of the basin and join the river further down. At Dhawaleswaram which is at about 63 km away in a straight line from the sea, the river divides into two principal branches namely the Vasishta Godavari to the west and the Gauthami Godavari to the east. The Gauthami branch of the Godavari receives drainage through two of the sections (Ralli and Dhawaleswaram) of the dam while the Vainateyam and Vasishta branches receive from the remaining two (Madduru and Vizzeshwaram) sections. The volume of water entering the Gauthami branch exceeds that which is drained into the Vainateyam and Vasishta braches. They flow through a wide delta to the sea. The Vasishta Godavari in its turn branches at Gannavaram into two, namely the Vasishta Godavari to the west and the Vainateyam Godavari to the east, the former opening into the Bay of Bengal at Antarvedi (Lat : 16°.18′ N, Long : 81°.42′ E) and the latter opening at Vadalarevu (Lat : 16°.25′ N, Long : 81°.58′ E). The Gauthami Godavari flows south east and opens into the Bay of Bengal at two places south of Yanam, namely at Kottapalem (Lat : 16°.35′ N, Long : 82°.17′ E) and at Bhiravapalem (Lat : 16°.41′ N, Long : 82°.23′ E). The Gauthami Godavari is also connected to the Kakinada Bay through two channels, the Coringa, a rising at Yanam and the Gaderu at Bhiravapalem. The Gaderu River was originally the principal drainage channel of Gauthami–Godavari and the Coringa River was also important for navigational and commercial purposes. In the course of the last hundred years, it had cut through the coast at two places i.e., near Kottapalem and Bhiravapalem villages. The confluence near Kottapalem is much older than that near Bhiravapalem. The Gaderu and Coringa rivers have since become narrow and shallow and gradually paled into insignificance. The Bay of Bengal pilot does not indicate the existence of the mouth of the Bhiravapalem and as such it is believed to be of recent origin.

The tidal portion extends up to 44 km in Gauthami, 42 km in Vasishta and 39 km in Vainateyam. Godavari is a typical positive estuary with a mean width of 1.5 km and an average depth of 12–15 km. The current systems prevailing in the estuary are mainly due to the tides. The waters rhythmiically played between the neritic and fresh water ends of the estuary.

Sampling and Analysis

Krishna Estuary: The water samples were collected from surface to bottom at three stations (Fig. 1) for two, low tides and two high tides during premonsoon (May'99), monsoon (Sep'99) and postmonsoon (Feb. 2000) seasons. Samples were collected from surface and bottom in Krishna main channel (7stations), Nadimeru (7 stations) and Gollumuttapaya (7 stations) during the dry season (April 2009) and wet season (Sep'2009) Water samples were collected by using water sampler designed by Ramkumar and Pattabhi Ramayya (1999)[5].

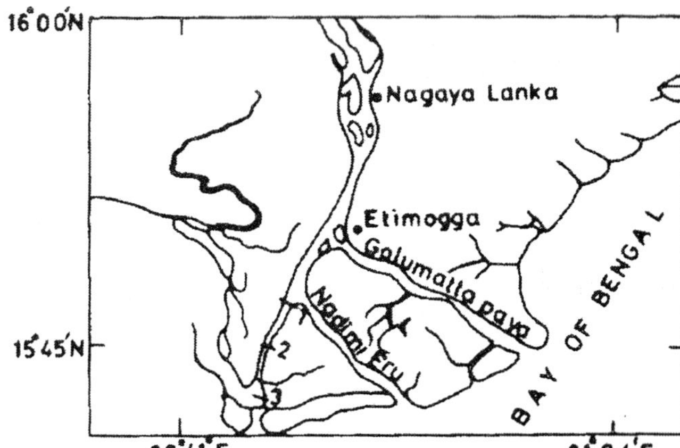

Fig. 1. Krishna estuary station locations.

Godavari estuary

The area of investigation and station locations are given in Fig. 2. The study area comprises of 3 coastal transects namely; Bhiravapalem, Vadalarevu and Antarvedi which covers the major coastal strip of Godavari river. On each transect 4 stations were fixed at 10, 20, 30 and 50 m depth contours, comprising of 12 coastal stations. Three estuarine regions were selected one each on the Gauthami (Between Bhiravapalem to Kotipalle), Vasishta (between Antarvedi to Rajole, Gannavaram), and on Vainateyam (between Vadalarevu to Lanka Gannavaram) rivers comprising of about 5 - 8 estuarine stations each to follow salinity-oxygen relationships in these estuarine regions. On the upstream 4 riverine stations were selected on Gauthami, Vasishta and Godavari (proper rivers) at Gannavaram, Jonnada, Siddantam and Dhawaleswaram, under very flood conditions during monsoon the number of stations in riverine and estuarine regions were reduced because of very high river discharges. Temperature, salinity and DO were measured immediately by Water Analyser Kit (Model : CMK-731, Century Instruments Ltd.), standardized with a sensitive thermometer (± 0.1°C) and standard chemical method for salinity and dissolved oxygen[6].

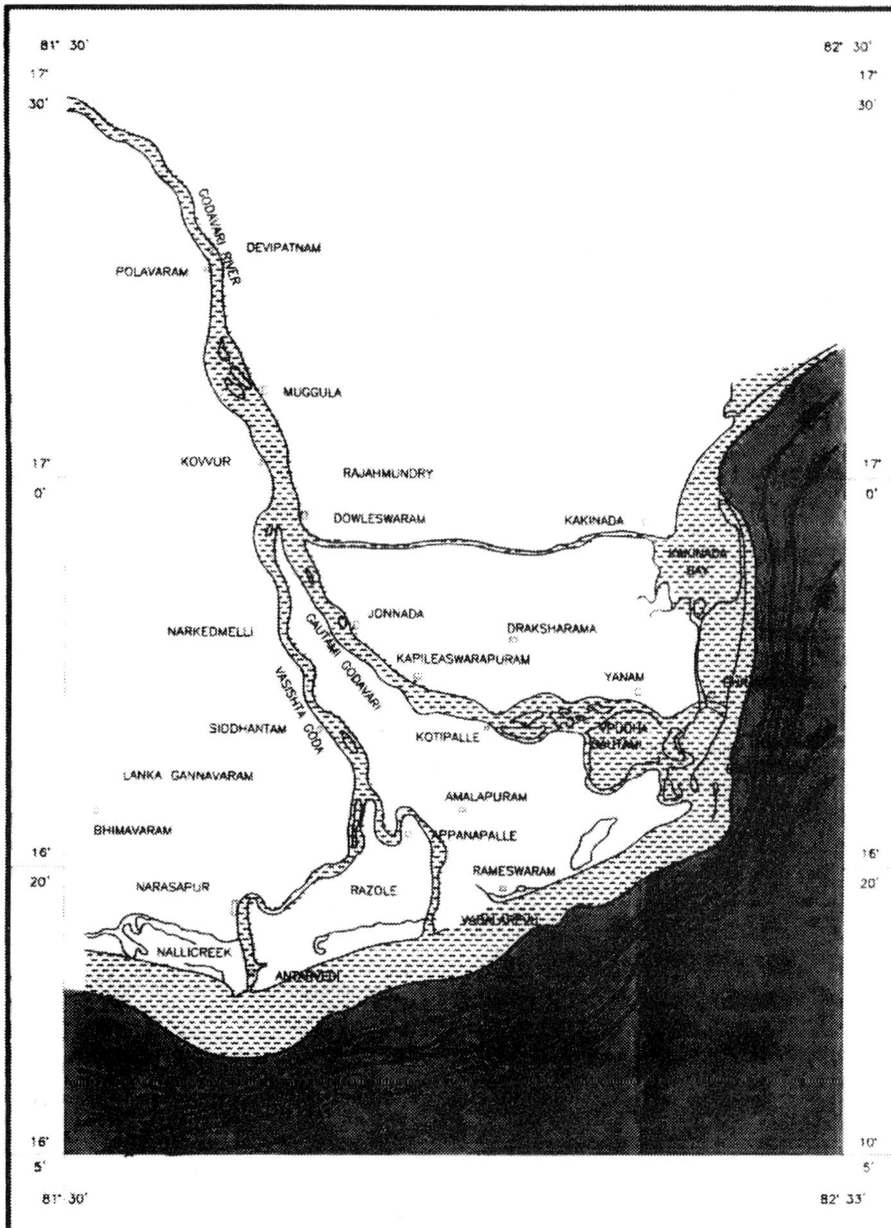

Fig. 2. Godavari estuary station locations.

Table 1: Some physicochemical parameters in Indian estuaries

Name of the estuary	Temperature(°C)	Salinity	Dissolved Oxygen (mg.dm^{-3})	Reference
	18 - 33	33.0 (summer)	6.84–7.03	De *et al.*, 1994[7]
	(MON)	0.07 (MON)	6.99 (MON)	Mukhopadhyay *et al.*, 2002[8]
Hooghly	(POM)	0.14 (POM)	6.56 (POM)	
	31.0 (PRM)	1.65 (PRM)	7.42 (PRM)	
			5.97-8.18 (EY)	Mukhopadhyay, 2007[9]
	22.5 – 31 (UR)	<0.5 (UR)	8.64 (POM)	Upadhyay, 1988.
	24.0-29.0 (LR)	8.0-32.0 (LR)	5.21 (M)	
Mahanadi	24.9-30.5 (2004-2005)	3.09-31.1 (2004-2005)	6.57-6.76 (2004-2005)	Subratnaik *et al.*, 2009[11].
	27.4-30.9 (2005-2006)	3.11-18.4 (2005-2006)	7.12-7.78 (2005-2006)	
	24.9-29.8 (2006-2007)	9.6-26.7 (2006-2007)	7.25-7.81 (2006-2007)	
	21.5 – 34.0	0.5-34.4 (S)	3.7-11.6 (S)	Saisastry & Chandramohan, 1990[35].
Godavari		0.5-34.8 (B)	3.1-9.6 (B)	
			6.09 – 9.66	Somayajulu, *et al.* 1993[36].
	27.2-33.5 (PRM)	0.3-34.3 (S)	6.13-8.71 (S)	Padmavathi, 1998[37].
Gauthami Godavari	26.6-29.3 (M)	0.3-34.3 (B)	3.2-8.46 (B)	
	23.2-25.6 (POM)			
			5.49-6.38 (MON)	Tripathy, *et al.*,2005[38].
	(DRY)	6.61-30.7 (DRY)	5.67-8.51 (DRY)	Rita, 2008[39]
	27.1-28.5 (WET)	0.21-2.15 (WET)	6.53-8.93 (WET)	
	30.5-31.4 (PRM)	0.3-29.0 (S)	6.19-7.80 (S)	
Vainateyam Godavari	28.4-31.9 (M)	0.3-30.6 (B)	5.04-7.31 (B)	Padmavathi, 1998[37].
	24.4-25.4 (POM)			
	(DRY)	9.01-30.7 (DRY)	4.8-7.07 (DRY)	Rita, 2008[39]
	27.6-28.0 (WET)	0.21-1.10 (WET)	6.5-9.2 (WET)	
	29.1-32.0 (PRM)	0.3-29.5 (S)	5.73-7.31 (S)	Padmavathi, 1998[37].
	27.9-31.2 (M)	0.3-28.9 (B)	4.34-7.80 (B)	
	23.6-25.0 (POM)			
Vasishta Godavari	(DRY)	0.90-32.4 (DRY)	2.99-6.79 (DRY)	Rita, 2008[39]
	27.8-28.2 (WET)	0.12-22.4 (WET)	5.63-9.21 (WET)	
	25.8 – 32.5	<0.5 – 31.6	4.0 – 7.7	Narappa *et al.*, 1982[40].
	29.0 – 31.6(PRM)	13.3 – 33.2 (PRM)	0.0-1.6(PRM)	
Krishna	28.3 – 30.3 (M)	2.6 – 12.4 (M)	0.7-7.9(M)	Present Study
	25.4 – 27.4(POM)	4.1 – 13.4 (POM)	3.7-9.0(POM)	
	(DRY)	0.25-34.7 (DRY)	0.1-5.76 (DRY)	Present Study
	26.0-37.3 (WET)	0.37-29.2 (WET)	1.59-6.96 (WET)	
	24.0 – 33.5 (S)	16.8 (backwater)	3.69 – 5.44	Krishnamurthy & Sundararaj, 1973[41].
	29.5 – 30.5	25.0 (mangrove)		
		10.4- 30.5		Ramanathan *et al.*, 1988[42].

Vellar	24.0 – 33.5 (1977) 26.5 – 33.5 (1978)	0.01-36.3 (1977) 0.27-36.5 (1978)	4.71-10.9 (1977) 5.64-10.5 (1978)	Chandran & Rammoorthi, 1984a[12].
	23.0-33.0	3.0-35.0	5.14-8.0	Ashok *et al.*, 2005[43]
	24.5-34.2 (2000-02)	15.3-35.5 (2000-02)	4.49-10.0	Sundaramanickam *et al.*, 2008[44].
Rushikulya	25.8 – 32.0 (1988) 23.5 – 33.2 (1989)	0.05-34.7 (S). 0.3-35.7 (B)	4.3-12.0 (S) 4.3-12.0 (B)	Gowda & Panigrahy, 1992[45].
Ashtamudi	26 – 35.5	0.12- 35.05	10.3 – 14.0	Nair *et al.*, 1983[13]. Nair&Azis, 1987[46].
Cochin Backwaters	23.0 – 32.0		0.5 – 6.0	Qasim *et al.*, 1969[47].
Cochin estuary	4.0-7.0	0.0-10.0	3.5 – 8.4	Arun, 2005[48].
	24.0-32.0	22.81-36.19	5.4 – 7.8	Murthy & Das, 1972[19].
Mandovi		0.05 – 34.5	4.3-7.63	Singbal, 1976[20].
			3.2-7.63 (S), 3.7-7.04 (B)	De Sousa & Sen Gupta, 1986[21]
	24-32.5	0.1-35.0	3.86-8.0	Padmavathi & Goswami, 1996[49]
		31.5-36.25,	5.26 – 6.91	De Sousa & Sen Gupta,
Zuari		0.13 – 33.3		1986[21].
	24.0-32.0	0.1-36.6	1.57-7.29	Padmavathi & Goswami, 1996[49]
Vembanad Lake	28.30-29.80 (PRM) 29.80-29.90 (M)	6.45-8.60 (PRM) 0 – 0.6 (M)	5.61-8.43 (PRM)	Devassy & Gopinathan, 1970[50].
Narmada	25.5 – 31.2 (FW) 27-29.7 (EW)	< 0.1 (FW) 0.2-31.63 (EW)	9.4-13.1 (FW) 9.3-11.9 (EW)	Zingde *et al.*, 1981[51]. Gajbhiye *et al.*, 1981[14].
Damanganga			3.8-6.3, 0.57-16.7 (M), 6.5-8.5	Zingde *et al.*, 1980[30].
Kolak			0.0-0.43	Zingde *et al.*, 1980a[52].
Par		20.9-35.3 (PRM) 4.0-34.0 (MON)	0.0-1.57	Zingde *et al.*, 1979[53].
Ambika		2.3-15.2 (May) 0.3-0.9 (October)	5.1-11.7 (LR) 0.0-5.43 (UR)	Vijayalakshmi *et al.*, 1981[16]
			1.8-5.5	Zingde *et al.*, 1985[54], 1986[55].
Auranga		23.04 (October) 35.25 (May)	0.0-3.2 7.29-11.4	Zingde *et al.*, 1985[55],
			2.8-12.4	Zingde & Desai 1987[56].
Purna			6.7-12.4	Vijayalakshmi *et al.*, 1981[16]
Mindola	4.72-33.82 (LR)	0.05-33.8	3.6-12.3	Vijayalakshmi *et al.*, 1981[16] Zingde *et al.*, 1988[57]
Bahuda		27.0-34.0 (PRM)	4.7-8.0	Sujatha Mishra *et al.*,1993[58]
Swarnamukhi		29.2-36.0	3.71-5.57	Reddi *et al.*, 1993[59].

Results and discussion

Review on the distribution of dissolved oxygen in Indian estuaries

Distribution of dissolved oxygen in major, medium and minor Indian estuaries has been studied extensively over the last 3 decades and the data along with those of temperature and salinity are given in Table 1. Temperature and salinity data indicate that they spreads around 25–32°C and 0–35 PSU respectively. Correspondingly the equilibrium dissolved oxygen at 0 PSU salinity and 25°C and 32°C temperatures are 5.79 and 5.11 mg/l. Similarly the equilibrium dissolved oxygen concentration in waters at 35 PSU salinity and 25°C and 32°C temperature corresponding to 5.64 and 5.12 mg/l respectively. Comparing the equilibrium dissolved oxygen concentration levels with the observed DO values in various estuaries, the following observations can be inferred.

- Dissolved oxygen in estuaries like Hooghly[7,8,9], Mahanadi[10,11] Vellar[12], Ashtamudi[13], Narmada[14,15] and Purna[16] indicated highly supersaturated conditions.
- The higher values in the ranges of most of the estuaries indicate significant supersaturated conditions in estuarine waters.
- Lower values in the ranges showed wide variation from nil concentration to again supersaturated conditions.
- Most of the estuaries exhibited significant seasonal variation in dissolved oxygen.
- As expected in monsoon or wet seasons, high dissolved oxygen values were recorded compared to post or premonsoon or dry season periods.
- Minor annual changes in dissolved oxygen were reported for very few estuaries namely Vellar and Mahanadi.
- Some estuaries like Krishna, Vellar, Kolak, Par, Auranga, Swarnamukhi exhibited very low values of dissolved oxygen indicating hypoxic or anoxic conditions.

Distribution of dissolved oxygen in Indian estuaries is controlled, in common with other chemical parameters, by physical, topographic, biological and chemical processes. At the same time, it is very important to remember that no two estuaries are alike with respect to these influences.

- Tidal mixing of fresh and sea waters on semidiurnal scale in Indian estuaries with corresponding changes in the volume of water in an estuary produce temporal changes in contributions of dissolved oxygen in marine and fresh water sources.
- The circulation and especially the stratification of some estuaries leads to vertical and horizontal variations of dissolved oxygen. Mixing in Indian estuaries is season dependent with well mixed conditions in post

monsoon, partially mixed conditions in monsoon and stratified conditions in premonsoon periods[17].

- Estuarine topographic conditions that result in particularly restricted circulation leading to reduction in mixing of external and estuarine waters generate unusual chemical environments.
- The current regime in coastal and estuarine waters leads to the deposition or resuspension of various types of sedimentary materials influencing the distribution and budget of dissolved oxygen.
- One of the most important factors that control the distribution, addition and removal of dissolved oxygen in estuarine waters is the effect of temperature and salinity on its solubility.

Extensive studies on the distribution of dissolved oxygen in Hoogly[18] and Mandovi[19,20,21] considering the degree of undersaturation.

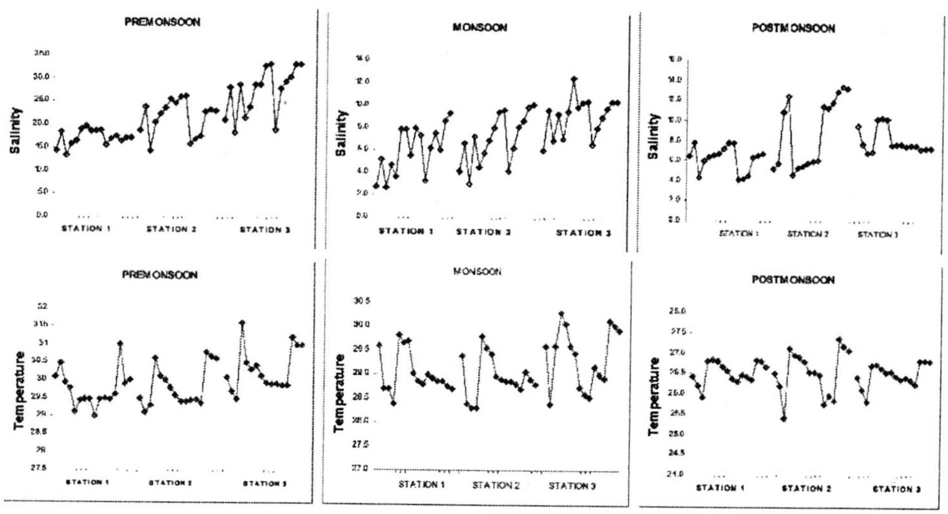

Fig. 3. Seasonal and spatial distribution of Salinity and Temperature in Krishna estuary.

Distribution of dissolved oxygen (DO) in Krishna estuary

DO values showed a decreasing trend from surface to bottom at stations 1 and 3 during premonsoon and at stations 2 and 3 during postmonsoon (Fig. 4, Table 2) But it showed an increasing trend from surface to bottom at stations 2 and 1 during premonsoon and postmonsoon respectively. Though no definite trend was observed in vertical variation during monsoon, it is in general, high in surface (0 and 0.2D depth) than in bottom waters, which may be due to the effect of photosynthetic activity of the phytoplankton and higher solubility of oxygen in low salinity water and high exchange rate at the air sea interface. The low DO values in bottom waters may be due to decrease in its solubility at higher salinity and consumption of oxygen due to decomposition of organic matter and respiration of bottom fauna and also may be due to rapid oxidation of organic

Table 2. Seasonal and spatial distribution of temperature, Salinity and dissolved oxygen in Krishna estuary.

Region	D	TEMPERATURE (°C)					SALINITY					DISSOLVED OXYGEN (ml/l)				
		1999-2000			2008-2009		1999-2000			2008-2009		1999-2000			2008-2009	
		PRM	MON	POM	DRY	WET	PRM	MON	POM	DRY	WET	PRM	MON	POM	DRY	WET
RIVER					27.9-30.0 (29.13)	26.0-27.1 (26.60)				0.25-14.68 (5.31)	0.37-0.61 (0.47)				3.36-5.76 (4.72)	6.24-6.96 (6.68)
ESTUARY																
Krishna	S	29.0-30.5 (29.72)	28.6-30.1 (29.26)	25.8-26.93 (26.46)	27.7-30.0 (28.93)	26.2-27.1 (26.68)	14.3-32.7 (21.33)	2.7-10.3 (6.67)	4.47-11.8 (7.27)	26.4-34.6 (31.1)	8.81-27.3 (19.33)	0.0-1.05 (0.51)	0.56-5.46 (2.41)	3.01-5.04 (4.04)	0.18-2.08 (0.95)	3.22-5.45 (4.54)
	B	29.1-31.6 (30.03)	28.3-30.3 (29.07)	25.4-27.36 (26.53)	27.2-29.2 (28.36)	26.0-27.0 (26.62)	13.3-33.2 (22.21)	2.6-12.43 (7.57)	4.06-13.4 (7.8)	26.6-34.7 (31.99)	10.66-28.0 (20.84)	0.0-0.91 (0.46)	0.49-5.06 (3.06)	2.57-6.32 (3.88)	0.15-1.99 (0.86)	2.66-5.24 (4.03)
Gollumutta paya	S				27.6-30.5 (28.75)	26.8-28.4 (27.33)				27.0-34.0 (30.71)	4.74-28.30 (16.21)				0.65-2.23 (1.31)	2.66-6.93 (5.00)
	B				27.6-28.7 (28.13)	26.4-28.2 (26.96)				29.1-34.3 (31.56)	5.25-29.2 (17.74)				0.44-1.86 (1.13)	1.59-6.55 (4.45)
Nadimeru	S				27.8-28.4 (28.09)	32.4-37.3 (35.36)				25.8-34.1 (31.14)	12.23-23.8 (17.73)				0.49-2.54 (1.57)	1.83-6.10 (4.73)
	B				27.1-28.0 (27.64)	32.0-37.0 (34.84)				28.2-34.2 (32.23)	15.2-23.5 (18.69)				0.40-1.09 (0.89)	1.68-5.99 (4.30)

PRM-Premonsoon, MON-Monsoon, POM-Postmonsoon, DRY- Dry Season, WET- Wet Season, S- Surface, B-Bottom, D-Depth

Table 3: Distribution of calculated dissolved oxygen and oxygen saturation anamoly in Krishna estuary

Region	Depth	C2(ml/l)					ΔG				
		1999-2000			2008-2009		1999-2000			2008-2009	
		PRM	MON	POM	DRY	WET	PRM	MON	POM	DRY	WET
RIVER					5.24-5.49 (5.37)	5.57-5.68 (5.62)				-3.63- +4.97 (-12.22)	9.95- 25.04 (19.06)
ESTUARY											
Krishna	S	4.55-4.96 (4.77)	5.0-5.36 (5.22)	5.19-5.56 (5.42)	5.26-5.42 (5.34)	5.52-5.61 (5.55)	-100- -77.17 (-88.03)	-89.08-+4.4 (-53.90)	-45.07- -2.89 (-25.36)	-96.68- -61.05 (-82.17)	-42.10- -1.59 (-18.19)
	B	4.40-4.90 (4.67)	4.98-5.29 (5.14)	5.15-5.51 (5.38)	5.31-5.46 (5.38)	5.52-5.58 (5.55)	-100- -78.46 (-89.78)	-90.37- -1.94 (-40.49)	-52.41-+17.04 (-27.78)	-97.27- -62.78 (-83.89)	-52.33- -5.63 (-25.82)
Gollumuttapaya	S				5.22-5.43 (5.35)	5.38-5.56 (5.50)				-88.04- -58.45 (-75.44)	-50.59- +24.92 (-9.35)
	B				5.35-5.44 (5.40)	5.40-5.59 (5.53)				-91.9- -65.79 (-79.02)	-70.63- +17.13 (-19.78)
Nadimeru	S				5.37-5.44 (5.40)	4.73-5.08 (4.88)				-90.96- -53.28 (-70.90)	-61.93- +22.46 (-3.34)
	B				5.40-5.47 (5.43)	4.76-5.11 (4.91)				-92.68- -79.99 (-83.63)	-65.24- +17.06 (-12.96)

PRM-Premonsoon, MON-Monsoon, POM-Postmonsoon, DRY- Dry Season, WET- Wet Season, S- Surface, B-Bottom, D-Depth carbon in sediment and slow diffusion of dissolved gases[22]. Station 2 is recognized to be boosting turbidity maximum (TM) zone, which may also be a cause of thwarting photosynthetic activity there that results in higher DO values. Increasing trend of dissolved oxygen was observed from premonsoon to monsoon to postmonsoon, which may be interpreted in conjunction with the turbidity values that it is primarily due to light penetration and photosynthetic activities of microorganisms. During summer (premonsoon), in view of higher light penetration, the estuarine and shallow coastal waters become highly productive which results in consumption of available oxygen from the estuary. DO minimum occurs in summer under low fresh water discharge and higher temperature[23].

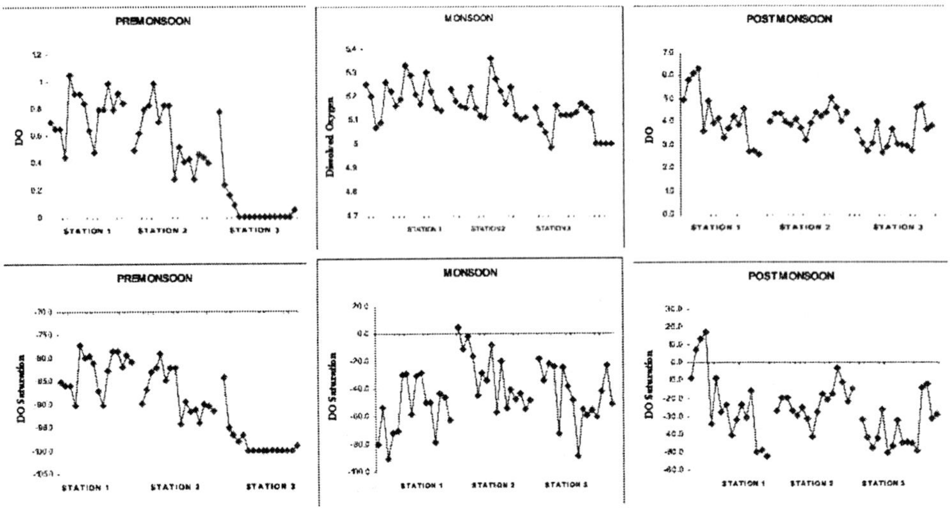

Fig. 4. Seasonal and spatial distribution of Dissolved Oxygen and DO Saturation anamoly in Krishna estuary.

No specific tidal impact is noticed in the estuary even though there is a wide seasonal variation. However, low values of DO were noted during low tide at stations 1 and 2 compared to high tide in pre-monsoon and monsoon seasons. Further It did not vary with tide in post-monsoon all along the estuary. Higher amounts of DO in waters in this region are generally regarded as the indication of healthy nature of the aqua system.

DO showed a decreasing trend in the order of postmonsoon > monsoon > premonsoon. This is due to winter cooling with low temperatures and high photosynthetic activity leading to high dissolved oxygen in postmonsoon (January/ February). Increased levels of DO in estuary inhibit denitrification resulting in release of dissolved nitrogen to the atmosphere[23] as well as increased nitrogen export to the sea[24] that may lead to increased productivity, introduction of exotic fauna and flora in coastal seas and resultant deterioration of water quality. High river inputs of fresh water with low salinity resulting in increase of oxygen solubility during monsoon (August-November) and low solubility as a result of high temperature and salinity during premonsoon (May). Padmavathi and Satyanarayana (1999)[25], while studying hydrographic characteristics of Godavari observed high dissolved oxygen in February followed by September and May. It was attributed to winter cooling with low temperature and high photosynthetic activity in January/February. Low dissolved oxygen content in May was explained on the basis of decrease in solubility with increase in temperature.

Dissolved oxygen, in general, decreased from head to mouth of the estuary in all the seasons and the variation is very much predominant in premonsoon compared to other seasons. This might be due to mixing of oxygen rich fresh water entered from the river and oxygen low saline water intruded from the sea as a result of existing circulation. However, Oxygen in monsoon exhibited an increasing trend from upper zone to middle zone and then decreased at mouth. Horizontal distribution is also affected by the amount of organic matter in both fresh and saline waters.

Hypoxic conditions in Krishna estuary

Oxygen solubility

Table 4 gives oxygen solubility (C”) values calculated using Weiss equation 1 and the percent oxygen saturation anomaly ΔG calculated by the equation

$$\Delta_G = \frac{C - C}{C''} \times 100$$

And the data are incorporated in Table 4.

Oxygen saturation concentration $[O_2]_{sat}$

The equilibrium concentration of oxygen with respect to the standard atmosphere at sea-level pressure is a function of the water temperature and salinity. The relation given below is taken from Benson and Krause (1984)[26]:

$$\ln[O_2]_{sat}\,(T,S) = \ln[O_2]_{sat}\,(T) + f(T)S$$

where, T refers to the absolute temperature[K] and S to the salinity [psu]. It includes a function $\ln[O_2]_{sat}$ (T) giving the value of the saturation concentration at zero salinity and a temperature function f. These two functions are detailed in the following table.

$\ln[O_2]_{sat}\,(T) = a_0 + a_1/T + a_2/T^2 + a_3/T^3 + a_4/T^4$	$a_0 = -1.3529996\ 10^2$ $a_1 = 1.572288\ 10^5$ $a_2 = -6.637149\ 10^7$ $a_3 = 1.243678\ 10^{10}$ $a_4 = -8.621061\ 10^{11}$
$f(T) = b_0 + b_1/T + b_2/T^2$	$b_0 = -2.0573\ 10^{-2}$ $b_1 = 1.2142\ 10^1$ $b_2 = -2.3631\ 10^3$

The concentration obtained by application of this relation is given in $\mu M\ O_2$. (Control value: for T = 273.15 and S = 35, $[O_2]_{sat}$ is equal to 347.88 μM or about 11.1 mg/l).

Table 4. Seasonal and spatial distribution of temperature, Salinity and dissolved oxygen in Godavari estuary

REGION	D	TEMPERATURE °C		SALINITY		DISSOLVED OXYGEN (ml/l)	
		1999-2001		1999-2001		1999-2001	
		DRY	WET	DRY	WET	DRY	WET
RIVER	S	28.2-28.7 (28.5)	27.9-28.0 (28.0)	0.48 (0.48)	0.21 (0.21)	6.8-6.9 (6.85)	6.45-6.57 (6.51)
	B						
ESTUARY							
Gauthami	S	27.7-29.0 (28.21)	27.5-28.5 (28.07)	7.47-30.72 (20.79)	0.21-2.15 (1.50)	4.17-5.96 (4.93)	4.57-5.60 (5.06)
	B	27.8-29.0 (28.41)	27.1-28.1 (27.67)	6.61-30.33 (19.80)	0.21-2.15 (1.50)	3.97-5.96 (4.88)	4.75-6.25 (5.41)
Vainateyam	S	26.0-28.9 (27.61)	27.8-28.0 (27.87)	9.01-30.64 (20.98)	0.21-1.10 (0.69)	3.97-4.95 (4.42)	4.57-6.45 (5.26)
	B	27.0-29.0 (28.1)	27.6	11.7-29.76 (22.05)	0.44	3.36-4.57 (4.10)	4.68
Vasishta	S	28.2-31.0 (29.73)	27.8-28.2 (28.07)	6.77-31.5 (21.53)	0.12-0.87 (0.62)	2.09-4.75 (3.93)	3.94-6.45 (4.78)
	B	28.0-30.8 (29.34)	28.00	0.90-32.37 (21.26)	22.38	3.47-4.57 (3.94)	4.17
COASTAL							
Bhairavapalem	S	27.2-28.4 (27.66)	27.0-27.8 (27.33)	27.16-32.8 (31.16)	0.87-33.74 (19.48)	4.06-5.15 (4.40)	2.09-3.74 (2.86)
	B	26.2-28.0 (26.97)	27.0-28.0 (27.38)	31.07-33.23 (32.61)	9.12-34.15 (27.34)	3.68-4.57 (4.15)	1.23-3.74 (2.53)
Vadalarevu	S	27.6-27.8 (27.73)	27.5-28.4 (27.98)	29.76-32.80 (31.72)	2.89-16.83 (12.99)	3.97-4.57 (4.29)	3.12-4.68 (4.09)
	B	26.2-27.0 (26.59)	26.0-28.2 (27.25)	23.65-34.02 (31.92)	17.17-18.90 (18.25)	3.77-4.75 (4.20)	4.17-4.77 (4.56)
Antarvedi	S	28.4-29.1 (28.83)	27.8-29.1 (28.38)	31.07-31.93 (31.44)	22.6-31.02 (28.40)	3.59-4.26 (3.98)	2.76-4.77 (3.76)
	B	25.3-28.0 (26.77)	26.0-28.0 (27.01)	31.93-33.67 (32.73)	22.81-34.01 (31.68)	2.38-4.06 (3.67)	2.91-4.77 (3.43)

DRY- Dry Season, WET- Wet Season, S- Surface, B-Bottom, D-Depth

Table 5. Distribution of calculated dissolved oxygen and oxygen saturation anamoly in Godavari estuary (Rita, 2008)

Region	D	C2(ml/l)		ΔG	
		DRY	WET	DRY	WET
RIVER	S		5.47-5.48 (5.48)		17.87- 19.85 (18.86)
	B				
ESTUARY					
Gauthami	S	4.64-5.25 (4.88)	5.37-5.52 (5.43)	-10.11- +13.61 (0.87)	-14.82- +1.43 (-6.90)
	B	4.63-5.26 (4.88)	5.40-5.56 (5.47)	-14.18- +13.26 (-0.12)	-12.56-+12.39 (-1.06)
Vainateyam	S	4.67-5.27 (4.91)	5.45-5.49 (5.47)	-21.17- +3.66 (-9.63)	-6.07- +17.46 (-3.95)
	B	4.59-5.21 (4.84)	5.50	-33.39- -4.73 (-15.16)	-14.97
Vasishta	S	4.53-5.08 (4.72)	5.43-5.49 (5.45)	-57.82- -4.59 (-16.42)	-27.47- +17.40 (-12.51)
	B	4.54-5.40 (4.77)	4.84	-28.05- -8.21 (-17.19)	-13.75
COASTAL					
Bhairavapalem	S	4.59-4.68 (4.63)	4.58-5.29 (4.92)	-13.19- +11.59 (-4.91)	-54.33—18.87 (-41.60)
	B	4.61-4.69 (4.65)	4.58-5.30 (4.76)	-20.12- -1.08 (-10.72)	-73.28—24.75 (-46.93)
Vadalarevu	S	4.59-4.66 (4.61)	4.99-5.35 (5.10)	-14.86- -0.48 (-7.03)	-41.71- -7.40 (-19.54)
	B	4.64-4.91 (4.69)	4.94-5.13 (5.01)	-19.80- -1.78 (-10.49)	-15.64- -3.85 (-9.11)
Antarvedi	S	4.51-4.57 (4.54)	4.54-4.81 (4.65)	-20.47- -5.69 (-12.36)	-42.56- +4.03 (-19.02)
	B	4.58-4.75 (4.66)	4.60-4.82 (4.67)	-49.89- -12.10 (-21.06)	-38.12- +3.07 (-26.55)

PRM-Premonsoon, MON-Monsoon, POM-Postmonsoon, DRY- Dry Season, WET- Wet Season, S- Surface, B-Bottom, D-Depth

The positive sign indicates super saturation while negative sign denotes undersaturation. It is evident from the observed and calculated data on oxygen solubility (Table 5) that the dissolved oxygen concentration in the estuarine waters are far less than the equilibrium concentrations indicating under saturated conditions. ΔG which denotes the deviations of oxygen content from the equilibrium concentration varied from –100 to –77.5 during premonsoon, from

–90.4 to +4.4 during monsoon and from –50.7 to +17.0 during postmonsoon. Thus ΔG values varied from supersaturated condition during postmonsoon period to complete depletion with no oxygen during premonsoon in spite of various physicochemical processes favorable for its replenishment. This clearly indicates that some anthropogenic factors must be responsible for oxygen depletion in addition to the prevailing natural physicochemical processes. During the dry season ΔG varied from –91.9 to -58.4, from –92.7 to –53.3 and from –97.2 to –61.1 in Gollumuttapaya, Nadimeru and Krishna estuary respectively. In wet season it varied from –70.6 to +24.9, in Gollumuttapaya from –65.2 to +22.5 in Nadimeru and from –52.3 to –1.59 in Krishna estuary.

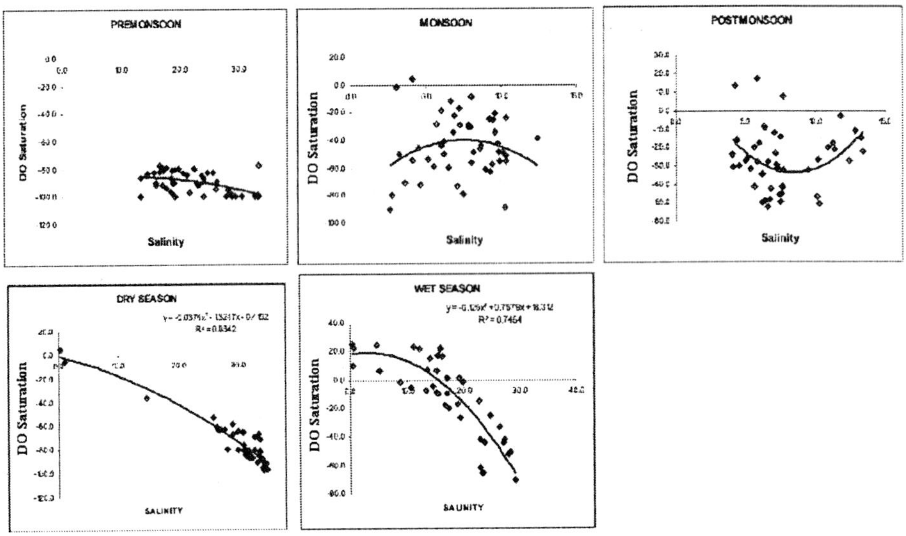

Fig. 5. Variation of ΔG with salinity in Krishna estuary

In Krishna estuary, prevailing undersaturated conditions with respect to dissolved oxygen can be due to high river run off with low dissolved oxygen content during monsoon with large amounts of decomposable matter with intermediate negative saturated anomaly values. Introduction of domestic sewage, agricultural and industrial effluents into the estuary may be a possible source of organic mater due to anthropogenic activity. Utilization of oxygen due to in-situ processes of high photosynthetic activity zone during premonsoon and postmonsoon seasons and due to sediment oxygen demand because of re-suspension of bottom sediments and bacterial activity at the bottom may be a factor affecting DO content. Another very important situation existing in the Krishna estuary is restricted water exchange conditions due to existing sand bar at the mouth and open by a shallow sill. Under such circumstances, complete removal of oxygen can occur. Similar situations were reported earlier[27,28,29]. Unsaturated conditions were reported by several workers in Indian estuaries such as Mandovi and Zuari[21], Hooghly[30], Damanganga[31] and other estuaries. Ghosh *et al.* (1987)[32] also reported under saturated conditions of incoming freshwater in the estuary during freshet. Further, undersaturated conditions in the Hooghly estuary[33,30] were attributed to large quantities of domestic sewage and utilization of dissolved oxygen in biochemical oxidation of organic matter brought from large catchment area during monsoon. All these observations in the behaviour of dissolved oxygen have been strongly supported by the data collected on oxidation reduction potentials measured in the Krishna estuary.

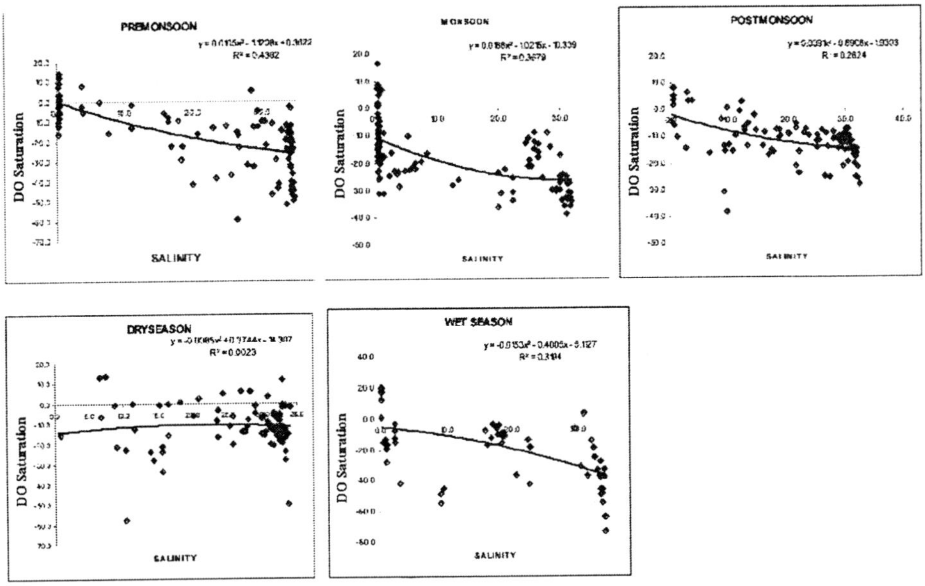

Fig. 6. Variation of ΔG with salinity in Godavari estuary

Krishna estuary with which huge activities of domestic, agricultural and industrial wastes are discharged has a well marked oxygen minimum. This is most pronounced under the low river discharge conditions which are generally associated with summer/premonsoon/ May.

Under low river discharge conditions tidal pumping introduces organic rich suspended material into the upper estuary. River water fully saturated with oxygen, high particle concentrations and relatively high temperatures of summer combine to provide ideal conditions for microbially mediated nitrification to ocean nitrification consumes oxygen and contributes to development of oxygen minimum as a result of the changes in dissolved oxygen.

Oxygen minimum may result due to other problems in the estuary.

1. Presence of a barrier to migratory fish species.
2. Small zones of anoxia in estuaries will lead to increased nitrogen exports to the coastal zone.
3. Higher DO concentration would inhibit denitrification, the process where by dissolved nitrogen species are removed to the atmosphere of change in estuarine oxygen status would also have implication for nitrification, the processes converting ammonia to nitrate.

Distribution of dissolved oxygen (DO) in Godavari estuary

During the dry season the dissolved oxygen varied from 3.97 to 5.96 ml/l (avg: 5.01 ml/l) in the estuarine region, from 3.86 to 5.15 ml/l (avg: 4.22 ml/l) in the coastal region of Gauthami. Dissolved oxygen varied from 3.36 to 5.76 ml/l (avg: 4.33 ml/l) in the estuary and from 3.77 to 6.34 ml/l (avg: 4.51 ml/l) in the coastal region of Vainateyam. And it varied from 2.09 to 4.75 ml/l (avg: 3.93 ml/l) in the estuary and from 2.38 to 4.26 ml/l (avg: 3.76 ml/l) in the coastal region of Vasishta Godavari. During wet season the dissolved oxygen varied from 6.45 to 6.57 ml/l (avg: 6.51 ml/l) in the river, from 4.57 to 6.25 ml/l (avg: 5.24 ml/l) in the estuary and from 1.23 to 3.74 ml/l (avg: 2.62 ml/l) in the coastal region of Gauthami. The dissolved oxygen varied from 4.57 to 6.45 ml/l (avg: 5.11 ml/l) in the estuary and from 3.12 to 4.77 ml/l (avg: 4.42 ml/l) in the coastal region of Vainateyam. And it varied from 3.94 to 6.45 ml/l (avg: 4.62 ml/l) in the estuary and from 2.76 to 4.77 ml/l (avg: 3.52 ml/l) in the coastal region of Vasishta.

DO decreased from surface to bottom in all the three stations during both the seasons, and also showed the decreasing trend from river > estuary > coastal during both the seasons and in all the three stations. It decreased from dry season to wet season in river and coastal areas but it showed the reverse trend in the estuary (Table 4).

ΔG which denotes the deviations of oxygen content from the equilibrium concentration varied from –14.18 to +15.95 in the estuarine region and from –16.56 to +11.59 in the coastal region of Gauthami Godavari during dry season, from –14.82 to +12.39 in the estuary and from –73.28 to –18.87 in the coastal region during wet season. In Vainateyam ΔG varied from –33.39 to +19.43 in the estuary and from –19.80 to +36.02 in the coastal region during dry season and it varied from –16.07 to + 17.46 in the estuary and from –41.71 to –4.50 in the coastal region during wet season. In Vasishta Godavari ΔG varied from –57.82 to –4.59 in the estuary and from –49.89 to –5.69 in the coastal area during dry season and from –27.47 to +17.40 in the estuarine region, from –42.56 to +4.03 in the coastal region during wet season.

The average percent oxygen saturation anomaly values indicate saturation conditions only in the Gauthami estuary during the dry season. However positive values in the estuary during the dry season indicate normal and healthy conditions. The negative average ΔG values in both Vainateyam and Vasishta estuaries during both the seasons indicate sub-oxic conditions during these months. Significant decreasing trend in ΔG values in the Gauthami (2.44, –3.98), Vainateyam (–10.91, –6.67) and Vasishta (–16.81, –12.82) estuarine waters observed during dry and wet seasons may be attributed to river discharge, tidal and stratification conditions. Contrary to normal trend of saturation conditions during wet season with high fresh water input, sub-oxic conditions during this season were attributed to high inflow of organic rich fresh water resulting in depletion of oxygen in estuarine waters. In general the coastal waters also exhibit sub-oxic conditions with negative ΔG values during both the seasons, although positive ΔG values were recorded ocassionally. ΔG values decreased from surface to bottom in the estuary in both the seasons.

ΔG values decreased from surface to bottom in the estuary in both the seasons. Negative values in estuary and coastal regions is indicative of suboxic conditions. However, positive values in the Gauthami estuary during dry season is indicative of normal and healthy conditions (Table 5).

Fluxes of dissolved oxygen

Water exchange

Assuming steady state conditions in the upper estuary, oxygen sources from water exchange with the river and the sea can be calculated using a model based on the conservation of salt (Balls *et al.*, 1996). The equation for mass balance can be written as:

$$R + Q_0 = Q_1 \tag{1}$$

where R is river flow, Q_0 is the flow of salt water into the upper estuary and Q_1 is the flow of brackish water out of the upper estuary.

From the conservation of salt:

$$Q_0.S_0 = Q_1.\ S_1 \tag{2}$$

where S_1 S_0 are the average salinities of water in the upper estuary and coastal water, respectively.

Rearranging the equations (1) and (2) gives:

$$R = Q_1\ .\ (1 - S_1/S_0) \tag{3}$$

Equation (3) was used to calculate Q_1 which was then substituted back into equation (1) to calculate Q_0. Both R and S_1 were obtained from Table 2 and S_0 was taken as 34.5. The model was then applied to DO and its concentrations in river and sea waters were taken as 7.24 and 6.77 mg. l^{-1} at water temperature of 25°C respectively. For each season, the net input of DO from water exchange can then be calculated from the equation. The corresponding data is included in Table 6.

$$R.O_{2R} + Q_0.\ O_{2S} = Q_1.\ O_{2E}$$

Table 6. Fluxes of dissolved oxygen in Krishna and Godavari estuaries

	water exchange (gr.sec^{-1})	**Gas exchange (gr.sec^{-1})**
KRISHNA		
Premonsoon	680.6	824.5
Monsoon	2547.8	538.8
Postmonsoon	58.9	187.9
ANNUAL	**3287.3**	**1551.2**
Dre season	258.2	663.2

Wet season	960.1	141.5
ANNUAL	**1218.3**	**804.7**
GAUTHAMI GODAVARI		
Premonsoon	39.3	-343.2
Monsoon	2999.6	475.2
Postmonsoon	-186.5	-246.8
ANNUAL	**2852.4**	**-114.8**
Dre season	783.1	-77.4
Wet season	7710.4	369.6
ANNUAL	**8493.5**	**292.2**
VAINATEYAM		
Premonsoon	75.8	24.6
Monsoon	4849.5	174.7
Postmonsoon	226.6	23.2
ANNUAL	**5151.9**	**222.5**
Dre season	994.8	252.7
Wet season	9878.2	114.1
ANNUAL	**10873**	**366.8**
VASISHTA		
Premonsoon	109	-176.4
Monsoon	4295.5	211.7
Postmonsoon	140	-117.8
ANNUAL	**4544.5**	**-82.5**
Dre season	1098.2	280.4
Wet season	12900	302.4
ANNUAL	**13998.2**	**582.8**

Gas transfer across the air-water interface

The flux (F) of atmospheric oxygen to the water column can be estimated from the equation:

$$F = k(C_s - C_{ob}).A \qquad (4)$$

where C_s and C_{ob} are respectively the saturated and observed DO concentrations in the estuary. A is the surface area of the estuary and k is transfer velocity.

The transfer velocity K can be calculated by using the equation Waninkhof (1992)[34].

$$k_{av} = 0.39\, u^2_{av}(Sc/660)^{-0.5}$$

u_{av} = average climatological wind speed (m/sec^{-1})

The Schmidt number for oxygen can be expressed as a function of temperature and salinity according to:

$$Sc = Sc_0 (1 + 3.14\ 10^{-3}S)$$

$$Sc_0 = 1800.6 - 120.1\ t + 3.7818\ t^2 - 0.047608\ t^3$$

where 'S' is the salinity in psu and 't' is the temperature in Celsius (Control value : Sc=590 at 20°C and 35 psu) and at 25°C Sc=464

The positive value of flux indicate transfer of gas from atmosphere to ocean that means undersaturated conditions in the system. The negative value indicate transfer of gas from ocean to atmosphere that means supersaturated conditions in the system.

Oxygen balance

The balance has been calculated by dividing the difference between the sources and sinks by the sum of all items in the budget, with the result expresses as a percentage. The difference in the balance may be due to denitrification and sediment water interactions in the estuary.

In Krishna estuary low oxygen waters occur during premonsoon season. The climate change may change the filter capacity of the ecosystem for dissolved oxygen in the future.

Acknowledgement

Author VRK is thankful to the authorities of the Delta Studies Institute, CSIR and DST New Delhi for providing financial support during the investigations.

References

1. Rao K L, India's Water Wealth: Its Assessment, Uses and Projections, Orient Longman Ltd., New Delhi, 1975 pp 267.
2. Dileep Kumar M, George M D, & Sen Gupta R, Inputs of Indian rivers to the Ocean : A synthesis. In : *Oceanography of the Indian Ocean*, B.N. Desai (Ed.), Oxford and IBH Pub. Ltd., New Delhi, 1992, pp 313 – 323.
3. Zingde M D, Environmental status of the coastal marine environment of India. *Management of Aquatic Ecosystems*, (1989) 37–57.
4. UNESCO, Discharge of selected rivers of the world. Vol. 2, monthly and annual discharges recorded at various selected stations, 1971 pp 194.
5. Ramkumar M & Pattabhi Ramayya M, Low cost water sampler for shallow water bodies *J. Geol. Soc. India*, 54 (1999) 93–96.
6. Grasshoff, K., 1983. Methods of sea water analysis, Verlag Chemie, Weinhein, pp 260.
7. De A K, Ghosh S K& Jana T K, Phytoplankton community organization and species diversity in the Hooghly Estuary, north east coast of India. *Indian J. Mar. Sci*, 23 (1994) 152–156.
8. Mukhopadhyay S K, Biswas H, De T K, Sen S & Jana T K, Seasonal effects on the air–water carbon dioxide exchange in the Hooghly estuary, NE coast of Bay of Bengal, India. *J. Environ. Monit.* 4(2002) 549–552.

9. Mukhopadhyay S K, The Hooghly estuarine system, NE coast of Bay of Bengal, India, Presented at the *Workshop on Indian estuaries*, NIO, Goa, 2007.

10. Upadhyay S, Physico-chemical characteristics of the mahanadi estuarine ecosystem, east coast of India. *Indian J. Mar. Sci*, 17 (1988) 19–23.

11. Subrat Naik, Acharya B C & Anil Mohapatra, Seasonal variations of phytoplankton in Mahanadi estuary, east coast of India. *Indian J. Mar. Sci*, 38 (2009) 184–190.

12. Chandran R & Ramamoorthi K, Hydrobiological studies in the gradient zone of Vellar estuary. I. Physico-chemical parameters. *Mahasagar-Bull. Natn. Inst. Oceanogr*, 17 (1984a) 67–77.

13. Nair N B, Dharmaraj P K A, Arunachalam M, Krishnakumar k & Balasubramanian N K, Ecology of Indian estuaries. Part 1- Physicoshemical features of water and sediment-nutrients of Ashtamudi estuary. *Indian J. Mar. Sci*, 12 (1983) 143–150.

14. Gajbhiye S N, Ram J, Nair V R & Desai B N, Plankton of the Narmada estuary and adjacent creeks. *Mahasagar- Bull. Natn. Inst. Oceanogr.*, 14 (1981) 23–31.

15. Zingde M D, Chander S, Rokade M A & Desai B.N, Base – line water quality of the river Narmada (Gujarat). *Indian J. Mar. Sci*, 10 (1981) 161–164.

16. Vijayalakshmi R Nair, Gajbhiye S N, Jiyalal Ram M & Desai B N, Biomass and composition of zooplankton in Auranga, Ambika, Purna and Mindola Estuaries of South Gujarat. *Indian J. Mar. Sci*, 10 (1981) 116–122.

17. Rajani Kumari, V., Studies on some hydrodynamic and hydro chemical characteristics of the Krishna estuary, east coast of India, Ph.D. thesis, Andhra University, Visakhapatnam, India, 2003.

18. Ghosh P B, Misra S, Choudary A & Ghosh A, Possible sources of aliphatic hydrocarbons in sediments of Hooghly estuary, north east coast of India. *Indian J. Mar. Sci*, 19 (1990) 145–147.

19. Murty C S & Das P K, Premonsoon tidal flow characteristics of Mandovi estuary. *Indian J. Mar. Sci.*, 1(1972) 148–151.

20. Singhbal S Y S, Diurnal variations of some physico-chemical factors in the Zuari estuary of Goa, *Indian J. Mar. Sci.*, 2 (1973) 90–93.

21. De Sousa S N & Sen Gupta R, Variations of dissolved oxygen in Mandovi and Zuari estuaries. *Indian J. Mar. Sci*, 15 (1986) 67–71.

22. Panigrahy P K, Das J, Das S N & Sahoo R K, Evaluation of the influence of various physico-chemical parameters on coastal water quality, around Orissa, by fator analysis. *Indian J. Mar. Sci.*, 28 (1999) 360–364.

23. Balls P W, Brockie N, Dobson J, & Johnston W, Dissolved oxygen and nitrification in the upper forth estuary during summer (1982–92): Patterns and trends. *Est. Coast. Shelf. Sci*, 42 (1996) 117–134.

24. Billen G, Somville m, De Becker E & Servais P, Nitrogen budget of the Scheldt hydrographical basin. *Netherlands Journal of Sea Research*, 19 (1985) 223–230.

25. Padmavathi D & Satyanarayana D, Distribution of nutrients and major elements in riverine, estuarine and adjoining coastal waters of Godavari, Bay of Bengal. *Indian J. Mar. Sci*, 28 (1999) 345–354.

26. Benson B b & Krause D Jr, The concentration and isotopic fractionation of oxygen dissolved in fresh water and seawater in equilibrium with the atmosphere. Limnol. Oceanog., 23, 3, (1984) 620–632.

27. Richards F A, In Chemical Oceanography Volume 2. 1st Edition. J.P. Riley and Skirrow (Eds.), Academic Press, London, 1965 pp 508.

28. Richards F A, Cline J D, Broenkow W W & Atkinson L P, Some consequences of the decomposition of organic matter in Lake Nitinat, an anoxic fjord. *Limnol. Oceanogr.*, Suppl, 10 (1965) 181–205.

29. Richards F A, Anderson J J & Cline J D, Chemical and Physical observations in Golfo Dulce, an anoxic basin on the Pacific coast of Costa Rica. *Limnol. Oceanogr*, 16 (1971) 43–50.

30. Ghosh S K, De T K, Choudary A & Jana T K, Oxygen difficiency in Hooghly estuary, east coast of India. *Indian J. Mar. Sci*, 20 (1991b) 216–217.

31. Zingde M D, Narvekar P V, Sharma R V & Desai B N, Water quality of the river Damanaganga (Gujarat). *Indian J. Mar. Sci*, 9 (1980b) 94–99.

32. Ghosh, S.K., Jana,T.K., Singh, B.N. and Choudary, A., 1987. *Mahasagar- Bull. Natn. Inst. Oceanogr*, 20 (1987) pp 155.

33. Sen Gupta R, Proc 9th International Symposium on Tropical Ecology, Banaras Hindu University, Varanasi, India. (1987) Abstract – 235.

34. Wanninkhof R, Relationship between wind speed and gas exchange over the Ocean. *Jl. Geophy. Res*, 97 (1992) 7373–7382.

35. Sai-Sastry A G R & Chandramohan P, Physico-chemical characteristics of Vasishta-Godavari Estuary, east coast of India : Pre-pollution status. *Indian J. Mar. Sci.*, 19 (1990) 42–46.

36. Somayajulu B L K, Martin J M, Eisma D, Thomas A J, Borole D V & Rao K S, Geochemical studies in the Godavari estuary, India. *Mar. Chem.*, 43 (1993) 83–93.

37. Padmavathi D, Studies on the distribution and behaviour of some chemical constituents in riverine, estuarine and coastal waters of Godavari (Bay of Bengal). Ph.D. Thesis, Andhra University, Visakhapatnam, 1998.

38. Tripathy S C, Ray A K Patra S & Sarma V V, Water quality assessment of Gautami – Godavari mangrove estuarine ecosystem of Andhra Pradesh, India during September 2001. *J. Earth. Syst. Sci.*, 114 No. 2 (2005) 185–190.

39. Rita G M, A study on the water quality of the Godavari estuary based on the distribution of dissolved oxygen. M.Phil, Andhra University, visakhapatnam. 2008.

40. Narappa K V, Rao M S & Rao M P,Comparision of foraminiferal assemblages from Godavari and Krishna River Estuaries. *Indian J. Mar. Sci.*, 11 (1982) 220–224.

41. Krishnamurthy K & Sundararaj V, A survey of environmental features in a section of the Vellar Coleroon estuarine sysyem, south India. *Mar. Biol.*, 23 (1973) 229–237.

42. Ramanathan A L, Subramanian V & Vaithyanathan P, Chemical and sediment characteristics of the upper reaches of Cauvery estuary, east coast of India. *Indian J. Mar. Sci.*, 17 (1988) 114–120.

43. Ashok Prabhu, Perumal P & Rajkumar M, Diversity of microzooplankton in Parangipettai coastal waters, south east coast of India. *J. Mar. Biol. Ass. India*, 47 (2005) 14–19.

44. Sundaramanickam A, Sivakumar T, Kumaran R, Ammaiappan V & Velappan R, A comparative study of Physico-Chemical investigation along Parangipettai and Cuddalore coast. *J. Envi. Sci. Tech*, 1 (2008) 1–10.

45. Gouda R & Panigrahy R C, Seasonal distribution and behaviour of silicate in the Rushikulya Estuary, east coast of India. *Indian J. Mar. Sci.*, 21 (1992) 111–115.

46. Nair N B & Azis P K A, Hydrobiology of Ashtamudi estuary–A tropical backwater system in Kerala. Proceedings of National Seminar on Estuarine Management, N.B. Nair (Ed.), National Committee on Science, technology and Environment, Government of Kerala, Trivandrum., (1987) 67–74.

47. Qasim S Z, Wallershaus S, Bhattathiri P M A & Abidi S A H, Organic production in a tropical estuary. *Proc. Indian Acad. Sci.*, 59B (1969) : 51–94.

48. Arun A U, Impact of artificial structures on biodiversity of estuaries: A case study from Cochin estuary with emphasis on clam beds. *Applied Ecol. Envi. Res.*, 4 (2005) 99–110.

49. Padmavathi G & Goswami S C, Zooplankton ecology in the Mandovi–Zuari estuarine system of Goa, West coast of India. *Indian J. Mar. Sci.*, 25 (1996) 268–273.

50. Zingde M D, Chander S, Rokade M A & Desai B N, Base-line water quality of the River Narmada (Gujarat). *Indian J. Mar. Sci.*, 10 (1981) 161–164.

51. Zingde M D, Sabnis M M, Mandalia A V & Desai B N, Effects of industrial waste disposal on the water quality of river Kolak. *Mahasagar-Bull. Natn. Inst. Oceanogr*, 13 (1980a) 99–110.

52. Zingde M D, Narvekar P V, Sharma R V & Desai B N, Water quality of the river Damanaganga (Gujarat). *Indian J. Mar. Sci*, 9 (1980b) 94–99.

53. Zingde M D, Sharma R V & Desai B.N, pollution in river Par and its abatement. *Indian J. Mar. Sci*, 8 (1979) 266–270.

54. Zingde M D, Sharma P & Sabnis M M, Physico–Chemical investigations in Auranga river estuary, *Mahasagar-Bull. Natn. Inst. Oceanogr*, 18 (1985) 307–321.

55. Zingde M D, Narvekar P V, Sharma P & Sabnis M M, Environmental studies of the Ambika and associated river estuaries. *Mar. Pollu. Bull*, 17 (1986) 267–274.

56. Zingde M D & Desai B N, Pollution of estuaries of Gujarat–An overview. In : Contribution in Marine Science. Sastyabdaporthi Felicitation Volume on 60th Birthday of Dr. S.Z Qasim. Published by National Institute of Oceanography. Goa. 1987, pp 245–267.

57. Zingde M D, Rokade M A, Sharma P, Mandalia A V & Desai B N, Pollution assessment and flushing characteristics of Mindhola River estuary. In: Threatened Habitats (eds.) V.P. Agarwal & L.D. Chaturvedi, Published by Society of Biosciences Muzaffarnagar. (1988) pp 225–239.

58. Sujatha Mishra, Panda D. & Panigrahy R C, Physico-chemical characteristics of the Bahuda estuary (Orissa) east coast of India. *Indian J. Mar. Sci.*, 22 (1993) 75–77.

59. Reddi K R, Jayaraju N, Suriyakumar & Sreenivas K, Tidal fluctuation in relation to certain physico–chemical properties in Swarnamukhi river estuary, east coast of India. *Indian J. Mar. Sci*, 22 (1993) 232–234.

Role of coal based thermal power plant in climate change

G.C. Kisku

Environmental Monitoring Section, Indian Institute of Toxicology Research Council of Scientific and Industrial Research, Lucknow-226 001, India
[E-mail : kiskugc1@rediffmail.com]

Introduction

Coal is the only natural resource and fossil fuel available in abundance in India. Consequently it is used widely as a thermal energy source and also as a major fuel for thermal power plants producing electricity. India is world's 6th largest energy consumer, accounting for 3.4% of global energy consumption. At presently, India has a total power installed capacity is 138251.63 MW and growing rapidly. Sector wise current installed capacity is 73492.38 MW by coal based thermal power plants which is 53.15% of total installed capacity, 14581.71 MW i.e. 10.54% by Gas based thermal power plants 1201.75 MW i.e. 0.069% by Diesel Oil based thermal power plants, 10175 MW i.e. 7.35% by Wind power, 34680.76 MW i.e. 25.08% by Hydro electric power and 4120 MW i.e. 2.9% nuclear power reactors.

Due to India's economic rise, the demand for energy has grown an average of 3.6% per annum over the past 30 years. The Government of India has to increase very rapidly the installed electricity generating capacity at least 200000 MW by 2012 to achieve the ambitious mission of "Power for all by 2012". Sufficient steady power supply is also a basic need to achieve GDP target growth rate of 8-10% and quality life for every individual. The objective of this paper is to understand the role of coal power plants in climatic change through exit hot flue gas, steam and waste heat expelled from cooling tower.

The worlds power demands are expected to rise 60% by 2030. With the worldwide total of active coal plants over 50000 and rising. The International Energy Agency (IEA) estimates that the fossil fuel will account for 85% of the energy market by 2030.

In a fossil fuel power plant the chemical energy stored in fossil fuel (such as coal, fuel oil, natural gas or oil shale) and oxygen of the air is converted successfully into thermal energy, mechanical energy and finally electrical energy for continuous use and distribution across a wide geographic area. Most thermal power stations in the world use fossil fuel, outnumbering nuclear, geothermal biomass or solar thermal plant. A simple word equation for this chemical reaction is:

Fuel + Oxygen = Heat + Carbon dioxide + Water vapour

Depending on the temperature and flame parameters during combustion, however, some of the nitrogen can be oxidized, producing nitrogen oxides.

India has about 210 billion coal reserves that are being mined at the rate of about 300 million tons/year ranking India as the third largest coal producing nation in the world. But the problems associated with the use of poor quality coal which has high ash content (40-45%) and thus low calorific value.

This results in higher coal consumption for each MW power produced in India. Further, most of the coal is located in the eastern parts of the country and requires transportation over long distance, mostly by trains which run on diesel. Majority of the coal is of anthracite, bituminous and sub bituminous type. Only in the Southern part of India are there some lignite (brown coal) deposits. Table 1 gives the wide range of parameters found in different coal sample.

Table 1. Ranges of percentage for composition of Indian coal

S.No.	Constituents	Percentage
1	Carbon	38-60
2	Volatile matter	21-36
3	Moisture Content	15-20
4	Silicon oxide (SIO_2)	45-63
5	Aluminum oxide (Al_2O_3)	15-36
6	Iron oxides (Fe_2O_3)	2-20
7	Calcium oxide (CaO)	Trace-12)
8	Magnesium oxide (MgO)	Trace-5
9	Potassium oxide	0.02-0.60
10	Sodium oxide (Na_2O)	0.01-0.40
11	Titanium oxide (TiO_2)	0.02-2.5
12	Ash	40-45
13	Sulphur (SO_3)	0.3-8.3
14	Phosphorus (P_2O_5)	0.5-1.09
15	Calorific value (kcal/kg)	2100-3700

Materials and methodology

There are 2 multi flue stack of the Rihand Super Thermal Power Plant. Boiler No. 1 & 2 are attached to one stack having a height of 225 m while the Boiler No 3 & 4 are attached to 275 m stack. All stack monitoring were carried out at ESP outlet ducts and at a height of 40 m form the ground floor.

Stack sample were collected from the sampling from the port in the respective stack using Stack Mentoring Kit (Model APM-615, Envirotech make,

New Delhi). Particulate matter was collected isokinetically in pre weighed micro glass fiber thimble and quantified gravimetrically. SO_2, NOx, O_2% CO, CO_2% CxHx, Poison index, Stack gas temperature were directly measured with the help of Automatic Flu Gas Analyzer (Model KM-9106, Germany Made).

Particulate pollutants — Fly ash

During the combustion of pulverized coal, the oxidation of carbon, nitrogen and sulfur in the organic coal matter produces CO_2, NO_x and SO_x together with water vapour. The inert matter decompose to give residual solid material that are found in bottom as (~20% of the original mineral matter) ash fly ash (~80 %). In a typical 500 MW thermal power plant, fly ash generation (34% ash content @ 100% power load factor) work out as 2550 tons/day. Although some coal ash can be recycled for building material, road embankment and other use.

Gaseous pollutants

The flue gas from combustion of the fossil fuel is discharged to the air, this contents CO_2 and water vapour, as well as other substance such as nitrogen, nitrogen oxides, sulfur oxides, and (in the case of coal-fired plats) fly ash and mercury. Since Indian coal are low in sulphur content (0.3-0.4%); SO_2 emission from coal based thermal power station for each MW are lower than elsewhere in the world. However due to enormous amount of coal burnt, SO_2 emission is huge viz. from 500 MW thermal power station SO_2 emission work out to 105 tons per day or 4375 kg/hr (with sulphur content 0.7% and 100% load factor) and at the same time it also emits NO_2 and CO_2 which work out 24 tons per day or 1000 kg/hr and 6250 tons/day respectively.

Fly ash contain carcinogenic organic compound like chlorophenol and chloro benzene in ppm levels and polyaromatic hydrocarbon, polychlorinated dibenzo dioxibns, polychlorinated dibenzo furans etc. in ppb levels.

Some studies in Singrauli area on Hg emission from thermal power plants further indicate that emission of Hg both inorganic and organic of coal based thermal power plant, is significant.

Trace elements

Various studies reported that many elements present in fly ash, including Be, C, Ca, Si, K, Li, Mn, Na, P, Pb, Co, Ni, Cd, Cr, Cu, Hg, As, Se, Bo, Mo, are shown to be concentrated on particle surfaces. He found that for larger particles only small fraction of given element was present whereas for smaller particles the concentration was more.

Potential of coal power plant to climate change

Coal is the dirtiest most carbon intensive of all fossil fuels. It is one of the leading contributors to climate change. Burning of coal releases huge amounts of solid waste residue, massive amounts of gaseous pollutants SO_2, including green house gases CO_2, NOx and water vapors.

Besides, these solid, liquid and gaseous pollutants, it also releases substantial amount of direct heat into the atmosphere either through hot flu gases (50–70°C) from chimney, diffusion of hot stream from various points or waste heat to atmosphere in the cooling water system.

The cooling water used to condense the steam in the condenser returns to its source without having been changed other than having been warmed up to 35 to 55°C which may create thermal shock when discharged into the receiving water body if the hot water is not tempered with cool raw water.

Fossil fueled power stations are major emitters of green house gases (GHG) which according to the consensus of scientific organization are a major contributor to the global warming observed over the last 100 years. Brown coal emits 3 times as much as GHG as Natural gas, black coal emits twice as much.

Electricity generation using carbon based fuels is reasonable for a large fraction of carbon dioxide (CO_2) emissions world wide and for 41% of U.S. anthropogenic CO_2 emission. Of fossil fuel, coal combustion in thermal power station resulting greater amount of CO_2 emission per unit of electricity generated (2249 lbs/MWh) while oil produce less 1672 lbs/MW h are 211 kg/GJ) and natural gas produced the least 1135 lbs/MWh or (143 kg/GJ).

The Intergovernmental Panel on Climatic Change (IPCC) states that CO_2 is a green house gas and that increased quantities within the atmosphere will "very likely" lead to higher average temperature on a global scale (global warming) concerns regarding the potential for such warming to change the global climate prompted IPCC recommendations calling for large cut to CO_2 emission worldwide.

The world organizations and international agency, like the IEA, are concerned about the environmental impact of burning fossil fuel and coal in particular. The combustion of coal contributes the most to acid rain and air pollution and has been connected with global warming and climatic change.

Results and discussion

The stack emission data and analytical results are given in Table 2. The concentrations (of particulate matter, SO_2, NOx ranged 73-133, 537-541, 332-336 mg/Nm^3 respectively. The ambient temperature ranged between 302 (29°C) - 304 (31°C) with an average value of 302.75 (29.75°C) while stack gas ranged from 409 (136°C) - 418 (145°C) with an average value of 413.75 (140.75°C). The hourly volumes of CO_2 released from stack No. 1, 2, 3 and 4 253258, 277481, 207066 and 204237 Nm^3/hr with an average value of 235510 Nm^3/hr respectively. The plant was operating in full swing with greater efficiency. The Plant Load Factor ranged from 512-520 MW average 516 MW (PLF ~ 116%).

The huge amount of flue gas exit from the tall stack contents much higher temperature than background ambient temperature hence the excess amount of heat directly released and mixed into that cool atmosphere according to the Thermodynamics law. The green house gases like CO_2 and NO_X and also steam i.e. water vapour coming out from the leakage as well as automatic release of steam from the plant may trap the substantial amount of heat energy and may be the significant cause of climatic change.

Table 2. stack emission data and analytical results

S. No.	Particulars		Unit-I	Unit-II	Unit-III	Unit-IV
1	Date of Sampling		05/01/2009			
2	Stack Attached to		ESP Outlet			
3	Stack Height from Ground (m)		225		275	
4	Type of Stack		Multi Flue			
5	Type of Duct		Circular			
6	Material of Construction of Duct		Mild Steel			
7	Height of Sampling Port from the Ground (m)		110			
8	Diameter of Duct (m)		7.0			
9	Duct Area (m^2)		38.48			
10	Diameter of Sampling Port (mm)		150.0			
11	Type of Fuel Used		Coal			
12	Quantity of Fuel Used (MT/hr)		320	300	315	307
13	Load during monitoring (MW)		518	520	514	512
14	Atmospheric Pressure (mm of Hg)		740	740	740	740
15	Ambient Temperature (K)		302	302	303	304
16	Stack Gas Temperature (K)		416	418	409	412
17	Flue Gas Exit Velocity (m/sec)		26.49	26.2	20.5	18.4
	Flue Gas Exit volume (Nm^3/hr)		2558165	2522558	2380070	2127478
18	Sampling Period (min)		30	30	30	30
19	Concentration of Pollutants mg/Nm^3	SPM	119	133	73	92
		SO_2	538	541	537	537
		NO_x	328	336	332	332
		O_2%	19.7	19.9	10.9	9.5
		CO_2%	9.9	11.0	8.7	9.6
		CxHy (ppm)	235	500	53	142
		CO (ppm)	97	79	3	11

Conclusions

Emissions may be reduced through more efficient and higher combustion temperature and through more efficient production of electricity within the cycle.

Carbon capture and storage (CCS) of emissions from coal fire power station is another alternative but the technology is still being developed and will increase the cost of fossil fuel based production of electricity. CCS may not be economically viable unless the price of emitting CO_2 to the atmosphere rises. Efforts exists to use carbon capture and storage of emissions but these are not expected to be available on a commercial scale economically viable basis by 2025. Carbon dioxide emission from coal giving rise to global warming is almost absent in nuclear power plant. Hence India must explore alternative energy through the nuclear energy or renewable energy like solar, wind as well as hydroelectric power which India has great potential especially in the Himalaya region. The rivers originated from Himalayas are not only perennial but also have high current and steep gradient which favours the generation of hydroelectricity.

Hence, Government of India is already initiated to reduce the adverse environmental and ecological impact of coal-fired power plant through installation of more nuclear reactors and hydro projects.

References

1. http://en.wikipendia.org/wiki/ElectricitysectorinIndia
2. http://en.wikipendia.org/wiki/FossilFuelpowerplant
3. http://www.energypulse.net/centers/articles/articledisplay.cfm
4. http://hyperphysics.phy-astr.gsu.edu/hbase/thermo/heat.html#c3
5. Mishra U C, Environmental impact of coal industry and themal power plants in India. J. Enviro. Radio., 72 (2004), 35-40.
6. Karla A and Kisku G C, Project Report on Environmental Air Pollution and Management of Typical Thermal Power Plant. (2005).
7. Use of ash in Agriculture-Grey to Green Revolution, NTPC Ltd., R & D Building Centre, Noida.

Impacts of jhuming in Manipur

L.S. Singh[1]*, A. Pariari*[1]*, N.J. Singh*[2] *& G. Shukla*[3]

[1]Department of Spices and Plantation Crops, Faculty of Horticulture
[2]Department of Entomology, Faculty of Agriculture
Bidhan Chandra Krishi Viswavidyalaya, Mohanpur, Nadia, West Bengal, India
[3]ICAR - RCER - RC, Plandu, Ranchi (Jharkhand), India
[E- mail: dranupariari@gmail.com]

Introduction

The State of Manipur is a small landlocked, hilly and mountainous state within the north eastern India. It has 22,327 sq. km of area, which constitutes 0.7 per cent of the total land surface of India. The state has a small valley area of about 1,843 sq. km, which is only 8 per cent of the total area of the state. The remaining 92 per cent is hills and mountains. Agriculture is the single largest source of livelihood of about 70 per cent of the population and it is also the mainstay of the state's economy. The cultivation of rice is the major agricultural activity in Manipur and the main food for the people. The agro-climatic conditions are also much suitable for the cultivation of rice. Rice also contributes a major share of household income. Besides this, horticultural crops like fruits and vegetables are also grown under jhum. Churachandpur with 34% of the total area covered by jhum was highest among the hill districts followed by Chandel (25%), Ukhrul (22%), Senapati (10%) and Tamenglong (9%) [1]. Increase in jhum area, the intensity of jhuming and decrease in the jhum cycle are the main cause which causes decline in the forest area and the lost of rich biodiversity, bringing it on the verge of becoming totally degraded[2]. The rapid rate of deforestation in the state besides has also been the most potent factor for converting the land into wastelands. The paper discusses the jhuming practices in Manipur and strategies to revive the deforested areas for achieving ecological sustainability.

Land use and cropping pattern

The land use or land cover patterns of the state have been classified into five categories, namely settlement, agricultural land, forest, water bodies and others. The last category includes rivers/streams, roads, water logged areas *etc*. There are sharp differences in land use patterns between the valley and the hill districts. In the valley, settlements account for more than 10% of their respective areas whereas for the hill districts settlements account for less than 1% of the area.

Forest land for all the hill districts was above 95% of the areas. In the hilly areas, shifting cultivation is widely practiced, with settled terrace farming in foothill or low sloppy areas, above the adjacent rivers and streams. Rice and maize are the important crops grown both in the hills and plains. Other crops like, potatoes, sweet potatoes, root crops, cotton, chillies, tobacco, *etc*., along with plantation crops like tea, coffee, rubber, sandalwood, *etc*. and a variety of fruits like oranges, pineapples, plums, passion fruit and bananas are also grown in the state.

Percentage area under jhum

The maximum area under jhum cultivation is practiced in Churachandpur and Chandel (accounting for 45 per cent of the total area under jhum) and the lowest in Tamenglong and Senapati district though Senapati district has the highest proportion of forest area degraded by jhum. Churachandpur district covers an area of about 4, 57,000 ha and is located at 92°59′ E and 23°55′ N. Nearly 84.6 per cent of the population depends on jhum cultivation (Table 1).

Traditional practices of jhuming and jhum cycle

Jhum cultivation is generally practiced in the following sequence

i. Selecting a forest patch having mild slope with proper sunshine and clearing the vegetation normally in December and January

ii. Burning of the herbs, shrubs and twigs and branches (slashed vegetation) in February and March

iii. Sowing of seeds by dibbling-cereals, vegetables and oil seeds in April-May

iv. Continuing cultivation for a few years

v. Abandoning the cultivated site and shifting to other forest sites

With reduction in jhum cycle from 5-10 years to 2-3 years, the land under shifting cultivation rapidly loses the fertility of the surface soil and the families start moving to other virgin areas.

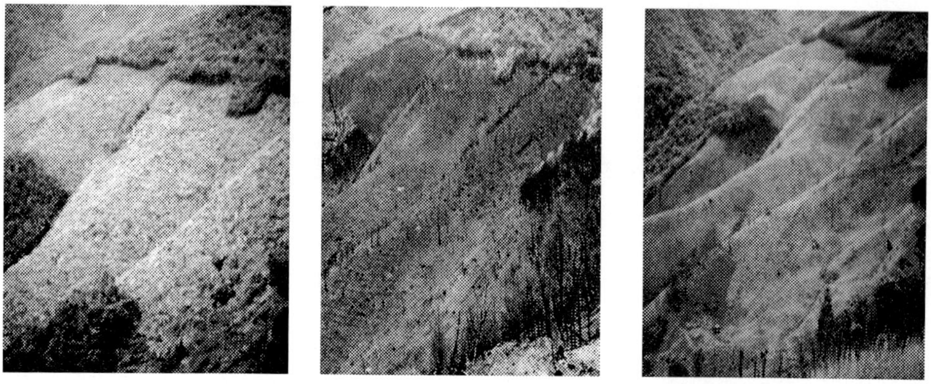

Jhuming cultivation area

Effect of jhuming on environment

Soil erosion and top soil degradation is the direct outcome of the clearing of natural vegetation. Slashing or felling down of trees, herbs and shrubs in jhum cultivation reducing oxygen generation, burning down of the sun-dried vegetation pollute the air with carbon-monoxide, nitrous oxides and many other harmful gases[4]. Various emissions from burning the fossil fuels including the dry vegetation contribute to the depletion of Ozone layer (a layer in the troposphere which protects the ultraviolet rays from the sun). In the last two decades the state has been witnessing various environmental degradations due to mass jhuming in the hilly areas. Although the impacts of jhuming are not felt by the masses because of lack of awareness. It is evident that jhuming is leading to a number of ecological catastrophes like frequent occurrence of landslide and mud slide in the hills causing major health hazards to the people of the state in particular and the North Eastern state as a whole. With the onset of rainy season, flash floods, mudslide, soil and land degradation and loss of fertile top soil are becoming regular happenings at every year. It is pertinent to mention here the incident of mudslide at Senapati District of Manipur in the month of July, 2004. There had been no such incident of this magnitude ever recorded in the history of Manipur that devastated lives of many people of the district[2]. It is of the view that it was the fall out of mass jhuming in the hill districts. The effect of jhumming cultivation can also be clearly seen from the quality and quantity of water from streams and rivulets. It is seen that the water streams jhuming area dries up earlier than years before. The scarcity of water during dry season is mainly caused by felling of big trees growing near and surrounding the streams and rivulets.

Effect on flora and fauna

Destruction of forest trees, flora and fauna, animals, birds and other living creatures in the forest are against the law of the Forest, Environment & Wild life Department, Ministry of the Government. Destruction of the natural habitats of these living organisms that brings ecological imbalance in the ecosystem is also forbidden by the forest laws and acts. Forest is the abode for many wild big and small animals and when jhuming cultivation takes place many animals are killed and others are driven out from their habitats causing great decrease in their population. It also indirectly affects the food chain in nature. This causes some rare species of animals and plants endangered. As an example, Horn bill and Slow loris which were once seen abundantly in the reserved forest of Khamlang village are now vanished and remain only in name for present generations.

Threatened species[5]

Trees	Orchids	Aromatic and Medicinal plants	Animals	Birds
Phoebe hainisiana	*Cymbidiumgi ganteum*	*Aquilaria agallocha*	Brow antler dear (Sangai)	Horn bil (Langmeidong)

Junglans regia	*Arundina graminifolia*	*Dioscorea deltoidea*	*Panthera tigris*	Hume's pleasent (Nong il)
Rhus hookeri	*Ascocentrum ampellecum*	*Rauwolffia serpentina*	*Hylobates hoolock*	
Ghrewia elastica	*Dendribium arachnites*		*Macaca speciosa*	
Flacourtia cataphracta	*Dendrobium densiflorum*			
Butea frondosa	*Dendrobium nobile*			
Aphanomixi polystachys	*Kalimpongia narajitii*			
Magnolia griffithii	*Paphiopedilum insigne*			
Cycas pectinata	*Paphiopedilum villosum*			
Podocarpus wallichii	*Spiranthes sp.,*			
	Vanda cuerulea.			

Abolishing jhum cultivation: Is it possible?

The destruction of natural vegetation under jhuming is alarmingly on the rise. The natural topography compels the people in the hills of Manipur to adopt the age old traditional cultivation which is the only means to sustain a living. The risks before the *jhumias* are to be considered before taking steps to ban it. The Government of India's policy to abolish the jhum cultivation and replacing it with growing of horticultural crops is in one sense, a good gesture towards the economic development of the people and saving the environment. However, the change of one's occupation over night is something which is not feasible on the part of the hill farmers. Provisions of funds and food to feed the farmers and their families till reaping the fruits of the newly adopted horticultural crops in lieu of the jhum cultivation would only motivate the farmers to switch over to the new methods. This is the reason why many farmers are not willing to change their profession of cultivation in the hills. Hence, a swift change over from shifting cultivation to any other forms of cultivation under the present conditions would be something impracticable for the poor farmers of the region (Table 2).

Role of the Government

The Govt. should be trust-worthy to the farmers. The scientific systems of cultivation should be imparted to the farmers and fulfill the promises which were laid before the farmers. Special incentives should be provided to the farmers of jhum to change their cultivation to a scientific method. Crop insurance should be implemented and food stock and essential items should be made available to the farmers who adopt the Government's policy of new agricultural technologies. The farmers should also be compensated properly against any natural calamity.

Replacing shifting cultivation and adopting an alternative cultivation system would then be a success.

Tips to the farmers

Farmers in the hill region should be aware of the new technologies emerging from time to time. Uneconomical practices of hill agriculture should be avoided while selecting the crops. Adequate cropping systems should be adopted to boost their income and maintain a sustainable self sufficiency in the household. Use of certain nutrients to replenish the vanishing food reserve in the soil should be adopted. The modern systems of plant protection and care should be given priority to harvest a bumper yield. Continuous cultivation of the selected spot can be cultivated continuously with amelioration and management of the soil. Soil amendments should be done where nutrient reserve is limited due to continuous cultivation through organic matters, mulching *etc.* The expertise of concerned authorities or personnel should be gained to tap the potential of the region.

Suggestion for improvement

i. Transfer of scientific technology from one tribe to another or one area to another.

ii. Upgrading of jhum by introducing variations in the species like cashew, coffee jackfruit, rubber and cinnamon which performs well in the region.

iii. Reduction of burning in the steep areas of hills. Stoppage of burning phase during shifting cultivation and substitution of shifting cultivation by development of bench terrace in land up to 50 per cent slope and contour bund in the low hill slopes for permanent cultivation and creation of irrigation facilities by water harvesting structures, channels, *etc.*

iv. Mulching should be done in the fields as soon as the crop season is over. This can be done by covering the fields using paddy stalks, corn-stalks, paddy husks, saw dust *etc.* so that the soil is not exposed and at the same time the soil gets enriched in organic matter.

v. Watershed management should be taken up properly in the foothills.

vi. Implementation of total literacy campaign.

vii. Providing employment opportunities and income generation on a regular basis through proper utilization of the land resources, *i.e.* by equitable distribution of waste land among the tribals.

viii. Encouraging cooperative efforts for carrying out forest-based activities, *i.e.* basket making, rope making, cane furniture processing of minor forest produce, honey collection, etc. have to be made commercially viable by providing proper marketing facilities. This will not only discourage tribals from practising shifting cultivation but will also help them financially.

ix. Formulation of Village Forest Committees for the protection and development of the degraded forests. Generating employment opportunities during the lean season of forestry operations. This will prevent tribals from shifting to other areas. Employing the tribals for collection of *kendu* leaves and *sal* seeds.

x. Agro-forestry can be effectively practiced on lands subjected to shifting cultivation. This will not only conserve the eroded land but helps in production of a forest crop and agricultural crop together.

Table 1. Area under jhum cultivation of some fruits and vegetables in Manipur (1994-95 to 2003-2004) ('000 ha)[3]

Crops	Pineapple	Orange	Lemon	Papaya	Banana	Bean	Cabbage	Cauliflower	Peas	Potato
1994-95	1.01	1.13	0.88	0.09	3.15	1.49	2.87	0.58	2.24	3.31
1995-96	0.57	1.06	0.90	0.10	3.07	1.40	3.03	0.88	1.82	3.41
1996-97	0.63	1.07	0.84	0.04	3.30	1.32	3.03	0.78	1.91	3.23
1997-98	0.46	0.99	0.67	0.06	4.41	0.60	3.87	0.76	4.45	3.23
1998-99	0.24	1.65	0.93	0.07	2.90	1.25	2.23	0.43	1.79	2.17
1999-00	0.30	1.66	0.82	0.09	3.51	1.72	2.28	0.44	2.28	2.47
2000-01	0.55	1.67	0.86	0.12	3.53	1.72	2.39	0.38	2.34	2.59
2001-02	0.45	2.11	0.92	0.02	4.42	1.66	2.71	0.40	2.44	3.04
2002-03	1.03	1.13	0.07	0.15	3.88	2.30	3.22	0.59	1.82	4.22
2003-04	0.73	1.71	0.37	0.07	4.14	2.75	6.63	0.76	1.23	4.98

Table 2. Jhum rice area as percentage of geographical area (1998 - 2002)[3]

District	1998	1999	2000	2001	2002
Senapati	2.89	2.89	2.89	2.89	2.72
Tamenglong	6.15	6.15	6.15	6.15	6.04
Churachandpur	6.42	6.42	6.53	6.42	6.39
Chandel	2.86	2.86	2.86	2.86	2.86
Ukhrul	1.86	1.86	1.86	1.85	1.86

Conclusion

Jhum control will only be possible if there are alternative land use systems. Alternative land use systems like horticulture with proper marketing and other infrastructural facilities will help the *jhumias* to divert the farming pattern. Until such systems are evolved, any attempt to stop jhum cultivation will amount to depriving the people of their basic means of survival. Afforestation in the hills and non agricultural activities should be given utmost priority with technical, financial and marketing supports from the Government and NGOs.

References

1. Brajakumar T, Manipur EMCB - ENVIS Node Newsletter (2010).
2. Dhamen T, Jhum cultivation major culprit behind extreme environment change in state, *Kangla Online*, (2010) 24th May.
3. Biren S, Land use and jhum cultivation in Manipur: problems, prospects and alternatives. *Ecology*, 28 (2005) 28-32.
4. Jamkhogin L, Jhum cultivation: Strategies for North East India. *Economic Development*, (2010) 12th January www.indigineousportal.com
5. Manihar B, Wasteland in Manipur: Problems and prospects. *Kurukshetra*, 57 (2009) 27-30.

A study of climate change due to global warming

Rama Mehta[1], Vipin Kumar[2] & Deepak[3]

[1]National Institute of Hydrology, Roorkee- 247667, Uttarakhand, India
[2]College of Engineering, Varthamanpuram, Roorkee- 247667, India
[3]Indian Institute of Technology, Roorkee- 247667, India
[E-mail: rama@nih.ernet.in]

Introduction

Climate change is a change in the statistical distribution of weather over periods of time that range from decades to millions of years. Climate change may be limited to a specific region, or may occur across the whole Earth. The warming of the climate system is unequivocal, as is now evident from observations of increases in global average air and ocean temperatures, widespread melting of snow and ice, and rising global mean sea level. The Earth's average surface temperature has risen by 0.76°C since 1850. Most of the warming that has occurred over the last 50 years is very likely to have been caused by human activities. Without further action to reduce greenhouse gas emissions, the global average surface temperature is likely to rise by a further 1.8-4.0°C this century, and by up to 6.4°C in the worst case scenario. Human activities that contribute to climate change include in particular the burning of fossil fuels, agriculture and land-use changes like deforestation. These cause emissions of carbon dioxide (CO_2), the main gas responsible for climate change, as well as of other 'greenhouse' gases. To bring climate change to a halt, global greenhouse gas emissions must be reduced significantly.

Greenhouse gases (for example, carbon dioxide, methane, nitrous oxide, water vapour, ozone), re-emit some of this heat to the earth's surface. If they did not perform this useful function, most of the heat energy would escape, leaving the earth cold (about –18°C) and unfit to support life. However, ever since the Industrial Revolution began about 150 years ago, man-made activities have added significant quantities of GHGs to the atmosphere. The atmospheric concentrations of carbon dioxide, methane, and nitrous oxide have grown by about 31%, 151% and 17%, respectively, between 1750 and 2000 (IPCC 2001).

Materials and methods

Global warming

The most commonly discussed measure of global warming is the trend in globally averaged temperature near the Earth's surface. This temperature rose by 0.74°C ±0.18°C over the period 1906-2005. The rate of warming over the last half of that period was almost double that for the period as a whole (0.13°C ±0.03°C per decade, versus 0.07°C ± 0.02°C per decade). Temperature is believed to have been relatively stable over the one or two thousand years before 1850, with regionally-varying fluctuations such as the Medieval Warm Period or the Little Ice Age.

Effects of global warming on climate change

The effects of global warming on climate change are of concern both for the environment and human life. Evidence of observed climate change includes the instrumental temperature record, rising sea levels, and decreased snow cover in the Northern Hemisphere. According to the IPCC fourth assessment report, "most of the observed increase in global average temperature since the mid-20th century is very likely due to the observed increase in greenhouse gas concentrations".

The effects of rising temperatures aren't waiting for some far-flung future. They're happening right now. Signs are appearing all over, and some of them are surprising. The heat is not only melting glaciers and sea ice; it's also shifting precipitation patterns and setting animals on the move. Some impacts from increasing temperatures are already happening. Some effects are as following:

- Ice is melting worldwide, especially at the Earth's poles. This includes mountain glaciers, ice sheets covering West Antarctica and Greenland, and Arctic sea ice.
- Excluding the ice caps and ice sheets of the Arctic and Antarctic, the total surface area of glaciers worldwide has been decreased by 50% since the end of the 19th century. Currently glacier retreat rates and mass balance losses have been increasing in the Andes, Alps, Pyrenees, Himalayas, Rocky Mountains and North Cascades.
- Global warming is projected to have a number of effects on the oceans. Ongoing effects include rising sea levels due to thermal expansion and melting of glaciers and ice sheets, and warming of the ocean surface, leading to increased temperature stratification.
- Sea level rise became faster over the last century. With increasing average global temperature, the water in the oceans expands in volume, and additional water enters them which had previously been locked up on land in glaciers, for example, the Greenland and the Antarctic ice sheets.

- Precipitation (rain and snowfall) has increased across the globe, on average.
- If the atmosphere is warmed, the saturation vapor pressure increases, and the amount of water vapor in the atmosphere will tend to increase. Since water vapor is a greenhouse gas, the increase in water vapor content makes the atmosphere warm further; this warming causes the atmosphere to hold still more water vapor and so on until other processes stop the feedback loop. The result is a much larger greenhouse effect than that due to CO_2 alone.

Other effects could happen later in this century, if warming continues. Those are as following:

- Sea levels are expected to rise between 7 and 23 inches (18 and 59 centimeters) by the end of the century, and continued melting at the poles could add between 4 and 8 inches (10 to 20 centimeters).
- Hurricanes and other storms are likely to become stronger.
- Species that depend on one another may become out of system. For example, plants could bloom earlier than their pollinating insects become active.
- Floods and droughts will become more common. Rainfall in Ethiopia, where droughts are already common, could decline by 10 percent over the next 50 years.
- Less fresh water will be available.
- Ecosystems will change. Some species will move farther north or become more successful; others won't be able to move and could become extinct. Since the mid-1980s, with less ice on which to live and fish for food, polar bears have gotten considerably skinnier. If sea ice disappears, the polar bears will as well.

The analytical effect of temperature on climate can be measured or calculated by empirical formulas. Sometime it is very difficult to measure due to bad weather. There is a new data driven technique called soft computing technique is available since last two decade. The present study is carried out using new emerging technique such as neuro-fuzzy subtractive clustering. Little detail has been given further.

Fuzzy logic theory (FLT)

Fuzzy logic is another area of artificial intelligence that has been applied successfully in different applications[1-3]. The key idea about fuzzy logic theory is that it allows for something to be partly this and partly that, rather than having to be either all this or all that. The degree of "belongingness" to a set or category can be described numerically by a membership number between 0 and 1.

Fuzzy logic interpolates the input into a crisp set. Every element in the universe of discourse, *X*, either belongs to a set or does not belong to the set. For example, the element in the universe of discourse, *X*, belongs to the set *A* or does not belong to *A* can be represented by the function.

$$\mu_A(x) = \begin{cases} 1 \text{ iff } x \in A \\ 0 \text{ iff } x \notin A \end{cases}$$

The above function is also called the characteristic function. The output is 1 if the element, *x*, belongs to set *A*, and 0 if the element, *x*, does not belong to set *A*.

Fuzzy interference systems

Fuzzy inference is the process of formulating the mapping from a given input to an output using fuzzy logic[4]. The mapping then provides a basis from which decisions can be made, or patterns discerned. The process of fuzzy inference involves: Membership Functions, Logical Operations, and If-Then Rules. A general Fuzzy Inference System has been shown in Fig. 1.

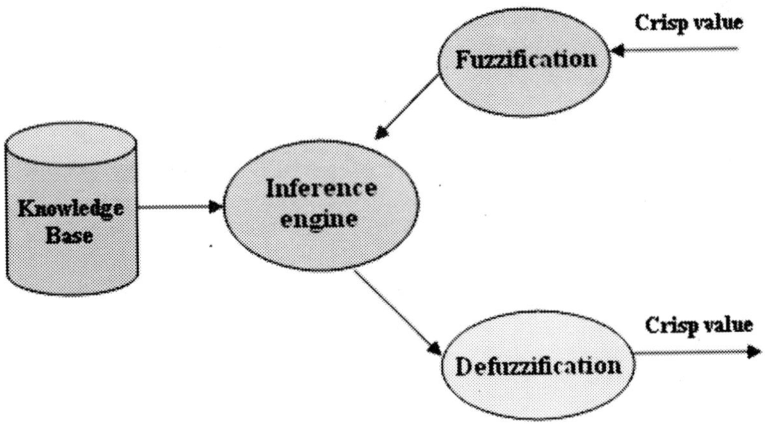

Fig. 1. A general Fuzzy Inference system

There are two types of fuzzy inference systems that can be implemented: Mamdani-type and Sugeno-type. These two types of inference systems vary somewhat in the way outputs are determined.

Neuro-fuzzy system

Neuro-fuzzy refers to combination of artificial neural networks and fuzzy logic. Neuro-fuzzy hybridization results in a hybrid intelligent system that synergizes these two techniques by combining the human-like reasoning style of fuzzy systems with the learning and connectionist structure of neural networks. Neuro-fuzzy hybridization is widely termed as Fuzzy Neural Network (FNN) or Neuro-Fuzzy System (NFS). Neuro-fuzzy system incorporates the human-like reasoning of IF-THEN fuzzy rules. The strength of neuro-fuzzy systems involves two contradictory requirements in fuzzy modeling: interpretability versus accuracy.

Jang (1996) introduced a novel architecture and learning procedure for FIS that uses a neural network learning algorithm for constructing a set of fuzzy if-then rules with appropriate MFs from the input-output pairs. This procedure of developing a FIS using the framework of adaptive neural network is called an adaptive neuro fuzzy inference system (ANFIS).

ANFIS operates on Sugeno type systems therefore there are only two types of output MFs: constant and linear. In MATLAB, ANFIS editor GUI invoked using "anfisedit", and training (calibration) and checking (validation) data sets are loaded. Two optimization methods used for FIS training are Hybrid (mixed-least squares and backpropagation) and backpropagation. The error tolerance is used to create a training stopping criterion, which is related to the error size. The training will stop if the designated epoch number is reached or the error goal is achieved, whichever comes first.

The property of the ANFIS model separates the problem into two sub-problem:

(i) parameter identification of the consequent part on grid base;

(ii) appropriate partitioning of the space of interest by clustering.

Cluster partitioning

In many situations, Fuzzy rules are difficult to identify by manual inspection and therefore are derived from observed data using techniques collectively known as fuzzy clustering.

The basic purpose of fuzzy clustering is to identify natural grouping of the data from a large data set, producing a concise representation of system behaviour. Various methods have been developed in the literature, such as fuzzy C-means clustering, mountain clustering, Gaustafron-Kessel (GK) fuzzy clustering and subtractive clustering. The subtractive clustering method has been used for present study.

Subtractive clustering

Subtractive clustering method is an extension of the mountain clustering method, where data points are considered as the potential candidates for cluster centres. It is a data clustering technique wherein each data point belongs to a cluster to some degree that is specified by membership grade. Clustering of numerical data forms the basis of many classification and system modeling algorithms. The purpose of clustering is to identify natural groupings of data from a large data set to produce a concise representation of a system's behaviour. Sub clustering tool is used to generate the ANFIS with minimum number of rules. The rules partition themselves according to the fuzzy qualities associated with each of the data clusters.

This approach is used to determine the number of rules and antecedent membership functions by considering each cluster centre (D_i) as a fuzzy rule. In this approach each data point of a set of N data points $\{x_1, x_2, \ldots, x_n\}$ in a p-dimensional space is considered as the candidate for the cluster centres. After normalizing data points in each direction, a density measure at data point x_i is computed on the basis of its location with respect to other data points and expressed as:

$$D_i = \sum_{j=1}^{N} e \left[-\left(\frac{2}{r_a} \right) . (x_i - x_j)^2 \right]$$

where r_a is a positive constant called cluster radius.

A data point is considered as a cluster centre when more data points are closer to it. Therefore, the data point (x_1^*) with the highest density measure (D_1^*) is considered as first cluster centre. Now excluding the influence of the first cluster centre, the density measure of all other data points is recalculated as

$$D_i^* = D_i - D_1^* . \mu_i (x_i^*) \quad (1)$$

$$\mu (x_i^*) = e \left[- \frac{\| x_i - x_i^* \|^2}{(r_b / 2)} \right] \quad (2)$$

where $r_b (r_b > r_a)$ is a positive constant that results in a measurable reduction in density measures of neighbourhood data points so as to avoid closely spaced cluster centres.

After the density measure for each data point (Eqs. 1 and 2), the data point with the highest remaining density measure is obtained and set as the next cluster D_2^* and all of the density measures for data points are revised again. The process is repeated and the density measures of remaining data points after computation of *kth* cluster centre are revised by substituting the location (x_k^*) and density measure (D_k^*) of the *kth* cluster centre in Eq. (2). This process is stopped when a sufficient number of cluster centres are generated. A sophisticated stopping criterion for automatically determining the number of clusters was suggested by Chau. These cluster centres (D_i^*) can be reasonably used as the centres for the fuzzy rules' premise and antecedent membership function that describes the system behaviour.

fuzzy rules' premise and antecedent membership function that describes the system behaviour.

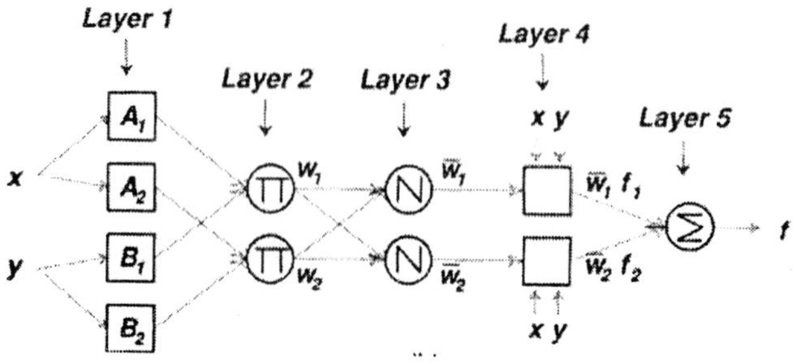

Fig. 2. General Architecture of ANFIS

The general architecture of the ANFIS is presented in Fig. 2. An adaptive network is a multi-layered feed forward structure whose overall output behaviour is determined by the value of a collection of modifiable parameters. Let us consider that the FIS has two inputs x and y and one output. For the first order Sugeno fuzzy model, a typical rule set with two fuzzy if then rules can be expressed as:

Rule 1: IF x is A_1 and y is B_1, THEN $f_1 = p_1x + q_1y + r_1$

Rule 2: IF x is A_2 and y is B_2, THEN $f_2 = p_2x + q_2y + r_2$

where A_1, A_2 and B_1, B_2 are the MFs for inputs x and y, respectively; p_1, q_1, r_1 and p_2, q_2, r_2 are the parameters of the input function.

Here the nodes of the layers have similar functions of Mamdani system except the linear or constant output.

Study area

In our study, we have considered the eastern part of Uttar Pradesh consisting of sixteen districts namely Allahabad, Azamgarh, Ballia, Chandauli, Deoria, Ghazipur, Gorakhpur, Jaunpur, Kushinagar, Maharajganj, Mau, Mirzapur, Sant Kabir Nagar, Sant Ravidas Nagar, Sonbhadra, and Varanasi (Fig. 3).

Data availability

We have considered six climate parameters: average temperature (TEMP), cloud cover (CC), potential evapotranspiration (PET), precipitation (PPT), vapor pressure (VP), wet day frequency (WDF). Out of these parameters, TEMP, CC, PPT, VP and WDF are input parameters and PET is output. We have taken data from the website **http://www.indiawaterportal.org/.** Monthly data has been taken for the 50 years period (from 1951 to 2000). The available data is partitioned in two parts. 60% values have been used for calibration/training the model and the rest 40% values were used for validation/checking of the model.

Twelve month-wise databases have been prepared for five inputs (TEMP, CC, PPT, VP and WDF) and single output (PET).

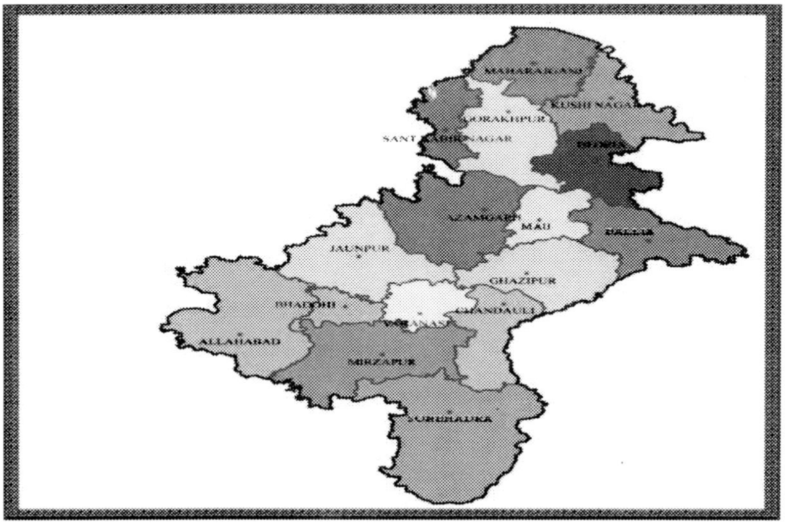

Fig. 3. Study area as Eastern part of Uttar Pradesh

MODEL DEVELOPMENT

The model has been developed using ANFIS editor of Fuzzy Logic Tool-Box in MATLAB

Table 1. Final parameters used in Anfis architecture

Error tolerance	0.001
Epochs	30
Squash factor	1.5
Reject ratio	0.15
Accept ratio	0.5
Range of influence	0.4

Table 2. Developed ANFIS Architecture for the Model

Number of nodes	116
Number of linear parameters	54
Total number of parameters	189
Number of training data pairs	360
Number of checking data pairs	240
Number of fuzzy rules	9

Training error	0.278518
Checking error	0.270824

Results and discussion

PET1 = Observed PET PET2 = Modeled PET

PET3 = PET when temperature is increased by 0.5°C

PET4 = PET when temperature is increased by 1°C

PET5 = PET when temperature is increased by 1.5°C

I1 = Percentage increment in PET when temperature is increased by 0.5°C

I2 = Percentage increment in PET when temperature is increased by 1°C

I3 = Percentage increment in PET when temperature is increased by 1.5°C

Performance of the model

The model performance is evaluated by using Nash-Sutcliffe criterion of percent variance explained (VAREX) as given below:

$$VAREX = \left[1 - \frac{\sum_{t=1}^{N} (O_t - P_t)^2}{\sum_{t=1}^{N} (O_t - \bar{O})^2} \right] \times 100$$

where N is number of observations; O_t is observed value at time t, Ō is mean of the observed values and P_t is predicted value at time t. The value of VAREX ranges from 0 (lowest performance) to 100 (highest performance)

Using above formula, we have

VAREX = 95.5

Some other performance evaluation indices from different methods (equations are given in appendix) are calculated in Table 3.

Table 3. Performance of the model using different criterias

Root-mean-square error RMSE	0.27
Mean absolute error MAE	0.22
Mean bias error MBE	0.01
Relative error R_e	3.42
Coefficient of efficiency R^2	0.95
Correlation CORR	0.97
Sum squared error SSE	17.60

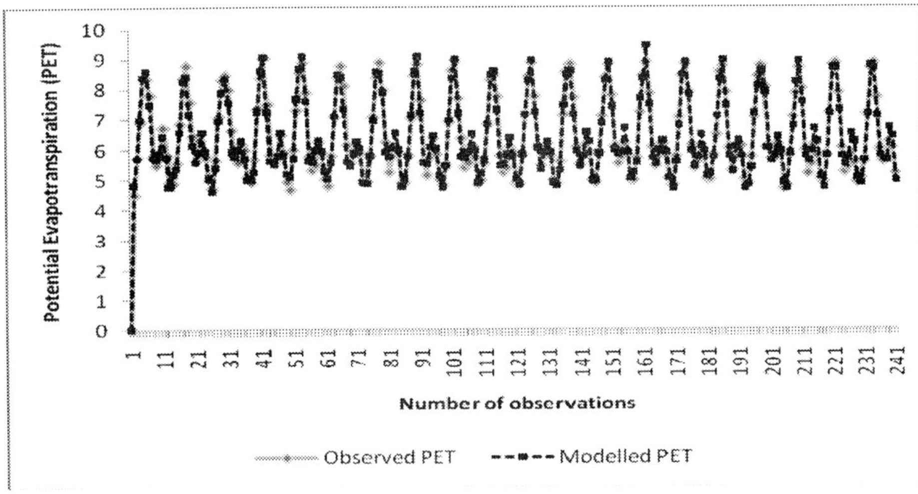

Fig. 3. Comparative plot between observed and modeled values of PET

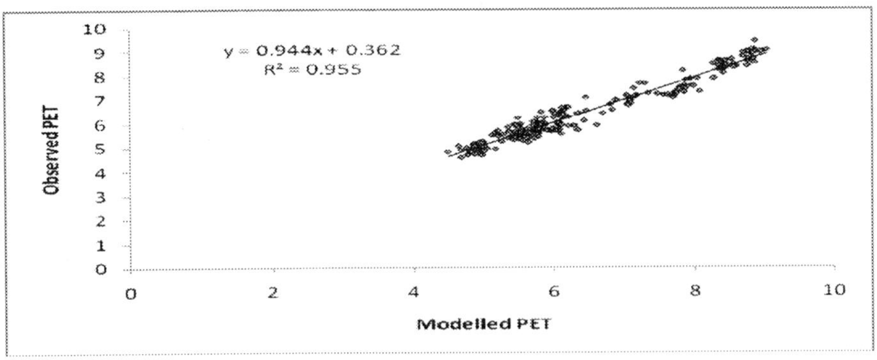

Fig. 4. Linear relationship between observed and modeled values of PET.

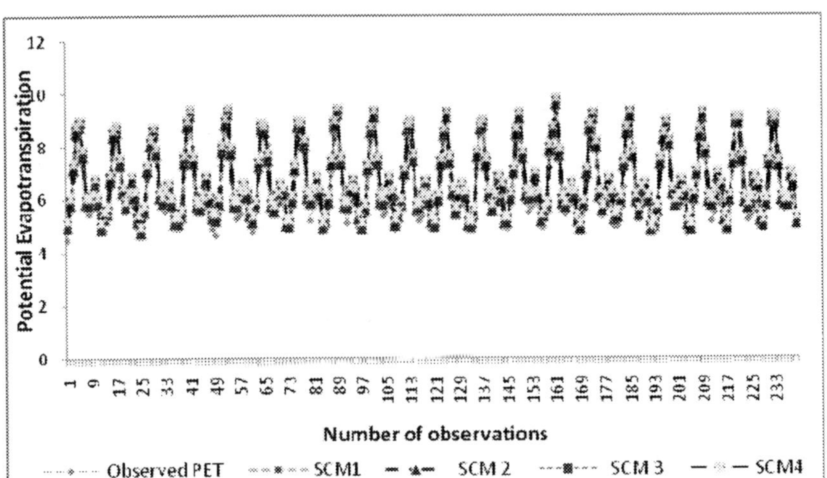

Fig. 5. Comparative graph for all PET values obtained with different increase in temperature while other four variables are fixed

Conclusions

Graphic User Interface (GUI) mode of the model is very easy to use for study of climate change in any region for any input data. It is clear that when temperature (TEMP) is increased by 0.5°C, 1°C and 1.5°C, potentail evapotranspiration (PET) of the region is increased by 4.05%, 5.33% and 7.06% respectively. Figures 1 to 4 show the consistent results. Figure 5 shows the increment in PET as temperature is rising. In the present study, PET has been taken as output and other five parameters (PPT, VP, WDF, CC & TEMP) as input data. But It is not necessary to take PET as output, one can take any other parameter (VP or WDF or CC) as output and remaining parameters as input data. Thus the effect of temperature increment (global warming) can be studied over the other climate parameters (VP, WDF or CC) also. The RMSE, MAE, MBE, R_e, R^2 and CORR are shown in Table 3. The efficiency of model is more than 95%. Anfis-cluster method gives better results with minimum error.

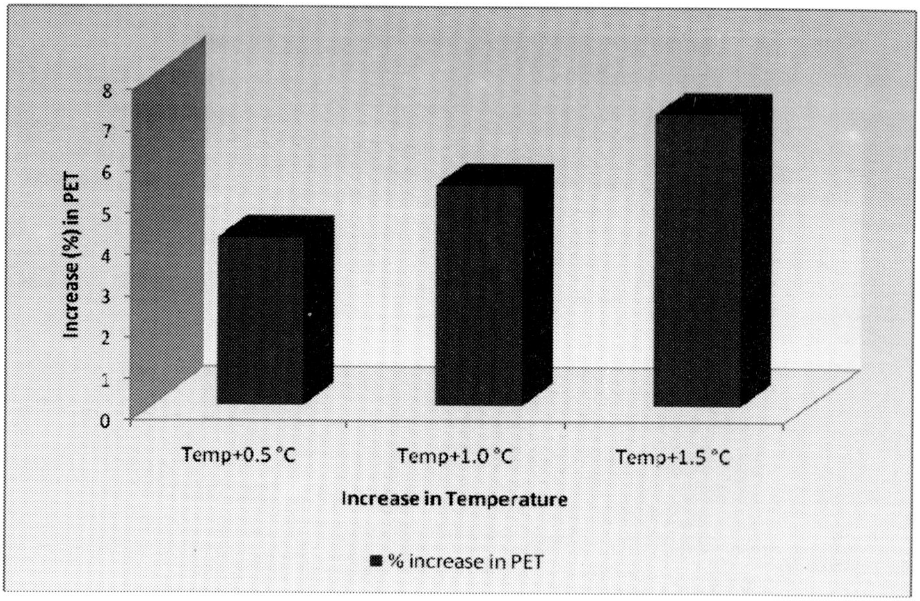

Fig. 6. Graph for variation in PET with increase in Temperature

Issues and recommendations

- As climate change is uncertain with time, space and quantity, Neuro-fuzzy technique is good as it depends upon the historical data.
- It may take considerable time to develop a data-driven model based on fuzzy technique. But once the model is developed, it usually runs faster than the conventional models.
- Models developed by Neuro-fuzzy technique are quite simple and easy to understand. The users or operators can easily use the developed models.

- The efficiency of a data driven model depends upon the range and quality of the source data. If the source data does not cover all possible state conditions, then the model may not perform well.

APPENDIX : Equations for performance evaluation indices

(1) Correlation CORR $= \dfrac{\sum_{t=1}^{n}\left(y_t^o - \overline{y^o}\right)\left(y_t^c - \overline{y^c}\right)}{\sqrt{\sum_{t=1}^{n}\left(y_t^o - \overline{y^o}\right)^2}\sqrt{\sum_{t=1}^{n}\left(y_t^c - \overline{y^c}\right)^2}}$

(2) Coefficient of efficiency $R^2 = 1 - \dfrac{\sum_{t=1}^{n}\left(y_t^o - y_t^c\right)^2}{\sum_{t=1}^{n}\left(y_t^o - \overline{y^o}\right)^2}$

(3) Root-mean-square error RMSE $= \sqrt{\dfrac{\sum_{t=1}^{n}\left(y_t^o - y_t^c\right)^2}{n}}$

(4) Sum squared error SSE $= \sum_{t=1}^{n}\left(y_t^o - y_t^c\right)^2$

(5) Mean bias error MBE $= \dfrac{1}{n}\sum_{t=1}^{n}\left(y_t^c - y_t^o\right)$

(6) Mean absolute error MAE $= \dfrac{\sum_{t=1}^{n}\left|\left(y_t^o - y_t^c\right)\right|}{n}$

(7) Relative error $R_e = \left|\dfrac{y_t^o - y_t^c}{y_t^c}\right| \times 100$

where y_t^o and y_t^c respectively are the observed and computed values at time t, $\overline{y^o}$ and $\overline{y^c}$ are the mean of the observed and computed values corresponding to n patterns.

References

1. Jang Jyh-Shing Roger, Neuro-Fuzzy and Soft Computing: A Computational Approach to Learning and Machine Intelligence, Prentice Hall of India, 1996, 335-367.
2. Klir George J. – Yuon Bo, *Fuzzy Sets and Fuzzy Logic*, Prentice Hall of India, 2008
3. MATLAB (R2007b), Fuzzy Logic Toolbox, The Mathworks, USA.
4. Zimmermann H J (1991) Fuzzy set theory and its application. Kluwer Academic, Dordrecht, The Netherlands.

Indian media coverage of climate change issues

I. Arul Aram

Department of Media Sciences, Anna University Chennai, Chennai-600 025, India
[E-mail: arulram@yahoo.com]

Introduction

The study is on challenges that journalists face while covering climate change. A qualitative approach has been taken in the methodology. The main objectives of study were to analyze stories related to climate change in English national dailies (The Hindu, The Times of India, The New Indian Express, Deccan Chronicle) and TV channels (particularly CNN IBN), to understand the gaps in effectively communicating information to the general public, to identify importance given to different climate change issues and to find out the problems faced by journalists while covering climate change and the factors that lead to the problems.

Materials and methods

The methodology used was discourse analysis and in-depth interview. Discourse analysis was done on stories from the national English dailies and the TV channel. The stories analysed include the Himalayan glacier error by the IPCC and the Copenhagen climate summit with respect to India's shifting grounds.

In-depth interviews were conducted with journalists from the four national English dailies and TV journalists, and qualitative results were drawn from the interviews. There was a consensus among the journalists that in this day and age climate change journalists have a responsibility towards making people aware of the brink they stand on and the stories should be pro-active.

The media has been quite successful in communicating the enormity of the risk planet earth is facing due to climate change. People have not been ignorant of climate change issue and are well aware of it. They agree that problems faced are due to human behaviour and not natural causes.

Results and discussion

Climate change journalism is plagued by the inability of journalists to look at macro issues. Most of the reports fail to link ground realities to existing policies and the politics behind such policies. Other challenges that plague climate change journalism in India are that the media organizations are not supportive of journalists. They may not carry the story itself, even if they do they do not support the journalist if he/she gets legal notices for it.

The problems faced by climate change journalists are similar to those in other beats. Lack of sensational content may cause reports to get sidelined. So, controversies such as an error in assessing melting of glaciers in the Himalayas are blown up. It is difficult to find sources and one cannot get concrete facts. With extensive competition from other media organisations covering the same news story, the media has taken up to approach the story from different angles to retain the news hungry public.

Lack of credible research data on grassroots issues is another factor. Lack of local scientific data and scientific measurement methods too poses a problem.

Investors think environment/climate change journalists are out to damage their business. They exercise too much guard. When a journalist covers environment over a period of time there is a lot of antipathy from polluters, to the point that they refuse to even talk to the journalist. Corporate giants like Jindals, Vedanta and Kemplar are impediments to journalists. They contribute heavily to advertising, have contacts in the highest hierarchical levels in the media organizations and run smear campaigns against the very character of the journalist if he/she has presented them in a negative light.

The media prefers to report certain issues if activists or NGOs take it up to go out of their way to report on them because at the end of the day every media company is a corporate, it wants to make money and needs sponsors.

Television has covered the climate change issue quite extensively and most of its features have been focused on how man made activities have affected the environment on a long term basis. Television and newspapers play a vital role in communicating climate change news and information.

Television has a huge impact on shaping opinion on climate change issues. TV news translates the complex scientific concepts to the lay audience in a more cohesive and understand manner compared to other media. Visuals themselves tell a story! Unlike the newspapers it even reaches the illiterate crowd. On television, images are stronger than the words. The words merely provide an interpretation for the on-screen action. Sometimes even the sequence of the words lends to an intended vagueness and ambiguity in certain areas.

By nature there is no consistency in television reporting. They do not take up a permanent stand on issues. They concentrate more on immediacy than accuracy or even balancing. The content is being moved more towards entertainment to be 'marketised' i.e. the text operates more on a market basis, with a primary profit motive. Because of this approach the audiences' attention is diverted from the social and political issue and more towards the drama of the issue.

For journalists new to the topic, climate change is complex, thus training journalists to cover and deal with this issue must be made a priority for media outlets. It may happen that reporters without proper skills to report knowledgeably on the process of science and scientific uncertainty may inadvertently mess up the issue.

The public themselves are apathetic to climate change news unless it is local in nature and directly affects them. Media reporting needs to focus more on how climate change will affect people and less on controversial arguments. In these days of sound-bite journalism, television journalists, in particular, need to take care when reporting on uncertainty regarding climate change. They need to be qualified enough to differentiate unfounded controversies from genuine scientific disagreements among qualified experts, and present them legitimately.

The media needs to provide more coverage of climate change impacts on many aspects of daily life, such as tourism, sports/recreation, local government, health, business, agriculture, forestry, and education.

Indian climate policy: Challenges and resolutions

Saad Ullah Khan

Department of Mass Communication, Aligarh Muslim University, Aligarh—202 002
[E-mail: saadgeneration84@yahoo.co.in]

Introduction

From the past two decades, India is regularly experiencing the heat of climate change. From India's northern tip of Kashmir, where rising temperatures are playing havoc with local tourism and agriculture; to Sunderbans in the east, where people, mangroves and tigers are suffering; to the rising waters of the Bay of Bengal; to India's western metropolis of Mumbai with its 0.5 degree climb in the local temperature in just the past few years; and to Kochi in the south, where rising temperatures are unsettling the local crop economy. Recently, a report published by World Bank suggests that the developing world will suffer 80% of the damage from climate change despite accounting for one third of green house gases in the atmosphere[1]. Though, the threat of climate change is recognized by each nation in one way or the other, still we lack an effective international agreement on how to deal with the concerned problem. In the case of the Montreal Protocol covering ozone-depleting substances, there was a wide consensus and effective action was mobilized quickly. Thus, an understanding of perceptions and positions of different countries makes it easier to explore possibilities of effective action.

From the macro point of view, in this research paper, India's perceptions on the problem of climate change and sustainable development is presented; the kind of negotiating positions that follow from these perceptions; the policies India has undertaken so far and finally India's possibilities for action that can help contain the threat of climate change.

Risk of lower agricultural production

The FCCC objective states that GHG concentrations should be stabilized at levels where food production is not threatened (UN, 1992). Thus, by examining the impact on agriculture of different climate change scenarios, one can get an idea of what is tolerable. Rosenzweig and Parry (1994) have estimated significant adverse impact on the agriculture of many developing countries. In a more detailed study of India, Kumar and Parikh (2001a and 2001b) examined the impact of climate change on agricultural crop yields, GDP and welfare.

Considering a range of equilibrium climate change scenarios which project a temperature rise of 2.5°C to 4.9°C for India, Kumar and Parikh (2001a) estimated that: (a) without considering the carbon dioxide fertilization effects yield losses for rice and wheat vary between 32 and 40%, and 41 and 52%, respectively; (b) GDP would drop by between 1.8 to 3.4%. Their study also showed that even with carbon fertilization effects, losses would be in the same direction but somewhat smaller. Using an alternative methodology Kumar and Parikh (2001a) showed that even with farm-level adaptations, the impacts of climate change on Indian agriculture would remain significant. They estimated that with a temperature change of +2°C and an accompanying precipitation change of +7%, farm level total net-revenue would fall by 9%, whereas with a temperature increase of +3.5°C and precipitation change of +15%, the fall in farm level total net-revenue would be nearly 25%. For a developing country, these are very large changes which can cause much human misery. From India's point of view, a 2°C increase would be clearly intolerable. Other developing countries may be even more vulnerable.

Changes in monsoon pattern

Climate change has an effect on the monsoons too. India is heavily dependent on the monsoon to meet its agricultural and water needs, and also for protecting and propagating its rich biodiversity. Subtle changes have already been noted in the monsoon rain patterns by scientists at IIT, Delhi. They also warn that India will experience a decline in summer rainfall by the 2050s, summer rainfall accounts for almost 70% of the total annual rainfall over India and is crucial to Indian agriculture. Relatively small climatic changes can cause large water resource problems, particularly in arid and semi-arid regions such as northwest India. This will have an impact on agriculture, drinking water and on generation of hydro-electric power.

Apart from monsoon rains, India uses perennial rivers, which originate and depend on glacial melt-water in the Hindukush and Himalayan ranges. Since the melting season coincides with the summer monsoon season, any intensification of the monsoon is likely to contribute to flood disasters in the Himalayan catchment. Rising temperatures will also contribute to the raising of snowline, reducing the capacity of this natural reservoir, and increasing the risk of flash floods during the wet season.Besides, experts are now expecting one of the worst floods in Kasmir's history as aresult of melting of Himalayan Glaciers[2].

Unsustainable consumption patterns

Unsustainable consumption patterns of the rich industrialized nations are responsible for the threat of climate change. Only 25% of the global population lives in these countries, but they emit more than 70% of the total global CO_2 emissions and consume 75 to 80% of many of the other resources of the world (Parikh *et al.*, 1991). In per capita terms, the disparities are also large: an Indian citizen emits less than 0.25 tones of carbon per year whereas a citizen of the USA, for example, emits more than 5.5 tones. Parikh *et al.* (1991) emphasizes the need for an equitable and efficient solution to climate change and suggests that efficiency can be obtained through a system of tradable emission quotas and equity through equal allocation of global environmental space to all human beings. These findings were well received by developing countries and are echoed in the UNFCCC, which recognized the rights of developing countries to economic development and also the "common but differentiated responsibilities" of different countries.

India's accelerating emissions

Although not an emitter historically, India currently has one of the fastest growing economies in the world. With a government target of 8% GDP to achieve developmental priorities[3], a share of one sixth of the global population, and changing consumption patterns, India's emissions are set to increase dramatically. Growing at an almost breakneck pace, and guzzling coal, gas and oil in large quantities[3,4], we are today, the fourth largest emitter of greenhouse gases worldwide. Although our per-capita emissions are among the lowest in the world, our growth rates imply that the past is no predictor of the future **(5).** The most recent IPCC report suggests that India will experience the greatest increase in energy and greenhouse gas emissions in the world if it sustains a high annual economic growth rate. The International energy Agency predicts that India will become the third largest emitter of greenhouse gases by as early as 2015.

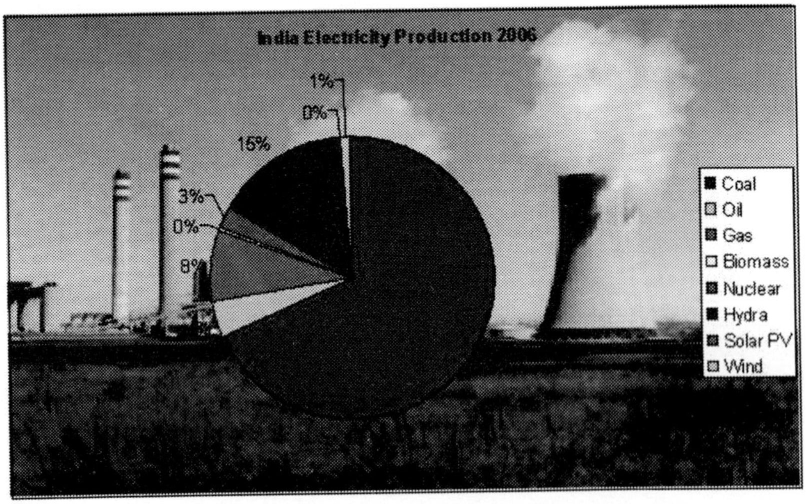

Fig. 1. India electricity production in 2006

India imports large quantities of fossil fuels to meet its energy needs, and the burning of fossil fuels alone accounts for 83% of India's carbon dioxide emissions[5]. Nearly 70% of our electricity supply comes from coal (Fig. 1). Although India has maintained its clear economic and social development imperatives, the government recognizes that climate change is a serious problem, and that business as usual is no longer the way forward.

Confronting climate change

India has for quite some time pursued GHG friendly policies in her own interest. India's obligations to minimize energy consumption — particularly oil consumption — and to deal with its environmental problems prompt it to follow many such policies. Directly or indirectly these efforts are made by Government as well as by people to reduce energy consumption. These include:

- Emphasis on energy conservation.
- Promotion of renewable energy sources.
- Abatement of air pollution.
- Afforestation and wasteland development.
- Economic reforms, subsidy removal and joint ventures in capital goods.
- Fuel substitution policies.

Some of these efforts are on-going for several decades and are institutionalized in a number of ways through policies, programs and the creation of specific institutions. These are government efforts; in addition there are a number of measures taken by people themselves. Some because of resource minimizing cultural traditions as well as good practices that exist in India and some due to sheer poverty and deprivation.

While some of the energy savings are due to conscious resource utilization practices in a positive sense, the dark side has to do with human drudgery and deprivation. These include "compulsory (or perverse?) forced energy savings" by the poor due to deprivation. These ranges from

(a) Lack of electricity connections and if connected, then a lack of electric appliances and even of adequate light bulbs in rural households.

(b) Lack of piped water or pumps that require long trips by women and children on foot to obtain surface water.

(c) Lack of even cooking fuels due to which the poor depend on biomass rather than clean and convenient fossil fuels.

(d) Lack of fans and heating devices for a large percentage of households that are necessary for comfort and productivity.

(e) Lack of basic infrastructure such as schools, hospitals, and roads that are essential elements for human development.

All of the above save energy at the cost of human welfare. Clearly, it is not recommended to continue the existing state of affairs. If India is committed to human development, poverty eradication should take place[6]. This may result in an increased energy use. This may be considered a due right of the poor, even if it increases India's GHG emissions.

Abatement of air pollution

Efficiency-enhancing measures in fossil fuel use may lead to the dual benefits of both local air pollution abatement and GHG emissions abatement. Pollution levels are prescribed for vehicles and have recently been made mandatory to reduce urban pollution. As a consequence, the transport sector is gearing up to face the challenges of urban air pollution by providing for more energy efficiency in vehicles. In addition, air pollution regulations result in higher fuel quality, which in turn lead to reduced GHG emissions. For example, improved gasoline can make vehicles run more efficiently[7]. It is difficult to say how much GHG can be saved due to such pollution control measures. Estimates indicate it to be in the range of 10 to 15% up to some point: say 2015 by which time the current stock would be largely replaced by the new vehicles. Then future reduction will be significant only if still stricter norms are followed at a later stage.

Afforestation and wasteland development

It is estimated that land-use change in developing countries could contribute to global emissions to the extent of 1.6 billion tones of carbon. However, indicators from India show that India's share of this contribution is minimal.

Biomass is widely used even today in India for a variety of purposes including fuel, timber and feedstock. Concern for the consequences-degradation of woods and forests and consequent degradation of soils was expressed as far back as 1974 (Fuel Policy Committee, 1974). More recently, programs for afforestation have found support from both governmental as well as non-governmental organizations.

These programs aim not only to halt deforestation but to increase green cover. Various agencies spend as much as Rs. 50 billion (US$ 1.25 billion) per year on wasteland development. If all the efforts at afforestation were to succeed, India's net emissions of CO_2 could come down significantly[7]. In the nineties, the rate of deforestation has slowed considerably, if not marginally reversed as shown by recent satellite imagery (MOEF, 1995).

Conclusions

An analysis of India's emissions shows that its per capita emission of carbon is one fourth of the global average. Even the top 10% of urban population emits well below the global average per capita emission. India realizes that its economy may be highly vulnerable to climate change. India also advocates the need of a common minimum policy in this region (**SAARC**), to counter the threats posed by the phenomenon of climate change. India's energy intensity in industry and transport sector has come down. It has installed 2300 MW of generating capacity based on various renewable. Deforestation is arrested and the vast potential of afforestation on wasteland is increasingly utilized. India and many developing countries have carried out price reforms and removed subsidies. These have resulted in substantial energy savings and reduction in emissions through greater use efficiency and fuel substitution. Developed countries are delaying action. By their delay, they are occupying global environmental space and are free riding on developing countries. The ***Clean Development Mechanism*** (CDM) could be risky for developing countries because of perverse incentives to exaggerate valid credits. We may start with fixing a global carbon price floor. A major attraction of the CDM for developing counties is technology transfer. However, carbon sequestration projects do not involve any significant technology. Developing country should be free to choose technology, not necessarily from the country that brings the CDM project, but from anywhere in the world. A scheme where a global trading system of carbon emissions with futures market is introduced.

References

1. World Bank, World Development Report 2010. Development and Climate Change (Washington D C ;World Bank, 2009).
2. Shafi Wani, Arif, "Valley's Erratic Climatic Change Worries Experts", *Greater Kashmir,* 5 October 2009.
3. Sharma S, Bhattacharya S and A Garg, Greenhouse gas emissions from India: A perspective. Current Science Volume 90, No. 3, p 326-333.
4. Imagining India: Nandan Nilekani.
5. Prasad and Raekha "India Builds a 2,500-mile Barrier to Rival the Great Wall of China" *The Times*, 28 December.
6. Chella Rajan and Sudhir, *Blue Alert-Climate Migrants in South Asia : Estimates and Solutions* (Bangalore : Greenpeace India Society, March 2008).
7. Mabey and Nick; Delivering Climate Security: International Security Responses to a Climate Changed World, Whitehall Paper 69 (London: Royal United Services Institute, 2009).
8. Military Advisory Board, National Security and the Threats of Climate Change (Alexandria, VA. The CNA Corporation, 2007) at www.securityandclimate.cna.org.

Hazard impact assessment due to severe storm using remote sensed data

[1]*Abhijat Abhyankar, Anand Patwardhan & Arun B. Inamdar*

Indian Institute of Technology, Bombay, Powai, Mumbai-400 076, Maharashtra, India
[1]National Institute of Construction management and Research,
Pune, India
[1][E-mail: abhijat1974@gmail.com]

Introduction

Remote sensing as a tool has been used frequently to map floods after the launch of earth observation satellites. After the successful launch of Landsat satellite series in late 1970s and early 1980s, new satellites were launched by various countries with similar configuration (in terms of bands and satellite parameters) like French launched Spot satellite series in 80s. India launched series of IRS satellites series. Other satellites launched during last decade include AVHRR, Quickbird, IKONOS and ASTER. All these satellites acquire digital images in visible, near IR and/or thermal IR bands of electromagnetic spectrum. The unavailability of optical images just after the cyclonic event results in under estimation of submergence and damage results. Hence, SAR data was used for monitoring floods due to its inherent advantage of cloud penetration capability and distinctive water response. Initially, Seasat and Shuttle Imaging Radar (SIR) showed the applicability of SAR data for floods monitoring. Later during last two decades new satellites with similar configuration were launched for similar applications (Radarsat, Envisat ASAR, JERS, ERS). The present study identifies submerged areas due to sever storm using satellite data.

The specific objectives of the present study are:

- to develop a framework for Hazard Impact assessment due to Cyclonic Storm using pre event IRS and SAR and post event SAR.
- to assess submerged area due to storm using remotely sensed data.

Materials and methods

Flood monitoring by satellites with optical data and microwave data have been reported in literature. Initially, researchers started using Landsat series to monitor floods[1-2]. The SPOT-5 and IKONOS satellites were used to monitor damage due to tsunami[3]. Indian Remote Sensing satellite series have been used to monitor damage due to floods/tropical cyclone[4]. The satellites with optical sensors data are unavailable during flood/cyclone due to cloud cover. This results in underestimated submergence results from satellites with optical sensors.

Due to unavailability of optical satellite data just after the cyclone, initially, Seasat and SIR data was used for monitoring floods[5-6]. New satellite platforms in recent two decade have been used for flood monitoring, namely Radarsat-1/2[7], Envisat ASAR[8], JERS[9], ERS-1/2[10-11]. It was reported that the water bodies have least backscattering values compared to other landcover classes[12-13].

Three approaches are reported in literature for damage assessment due to tropical cyclone/floods using remotely sensed data *i.e.* Approach one: uses pre and post event visible-NIR satellite images, approach two: uses pre and post event SAR images (with or without optical data as ancillary data), and approach three: use of pre event visible-NIR and pre-post event SAR image for submergence/inundation analysis. The present study developed a framework for damage assessment using pre event IRS and SAR and post event SAR and is depicted in Figure 1.

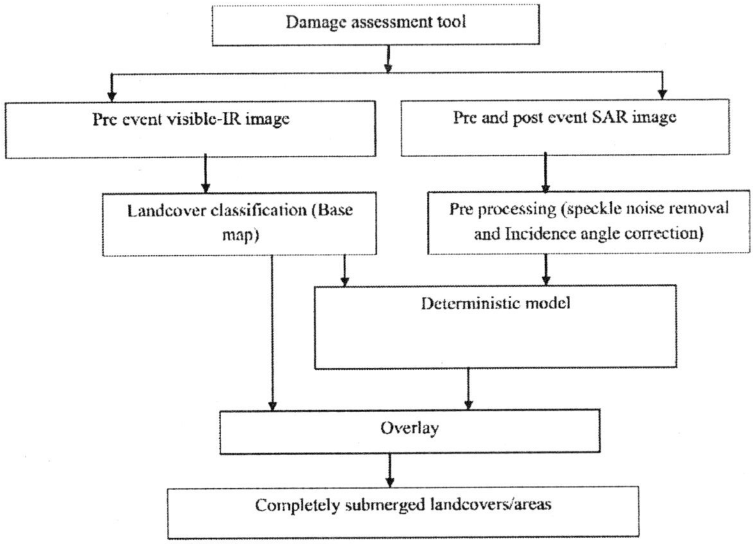

Fig. 1. Framework for Hazard Impact Assessment using pre event IRS and SAR and post event SAR

Study area

The Ogni storm crossed Andhra Pradesh Coast on October 30, 2006. It affected three coastal districts of Andhra Pradesh namely, Prakasam, Guntur and Krishna. In the present study, nine mandals of the Guntur district are considered for damage assessment namely, Tenali, Tsunder, Vemuru, Amruthalur, Cherukupalle, Pittalavanipalem, Karlapalem, Nagaram and Nizampatnam as they were reported to be impacted due to this storm. The total geographical area of nine mandals reported was 111 thousand hectares[14-15]. The map area using Geographic

Information Systems (GIS) was found to be 116.35 thousand hectares. The study area *i.e.* nine mandals considered for damage assessment of Guntur district due to this storm is given in Figure 2.

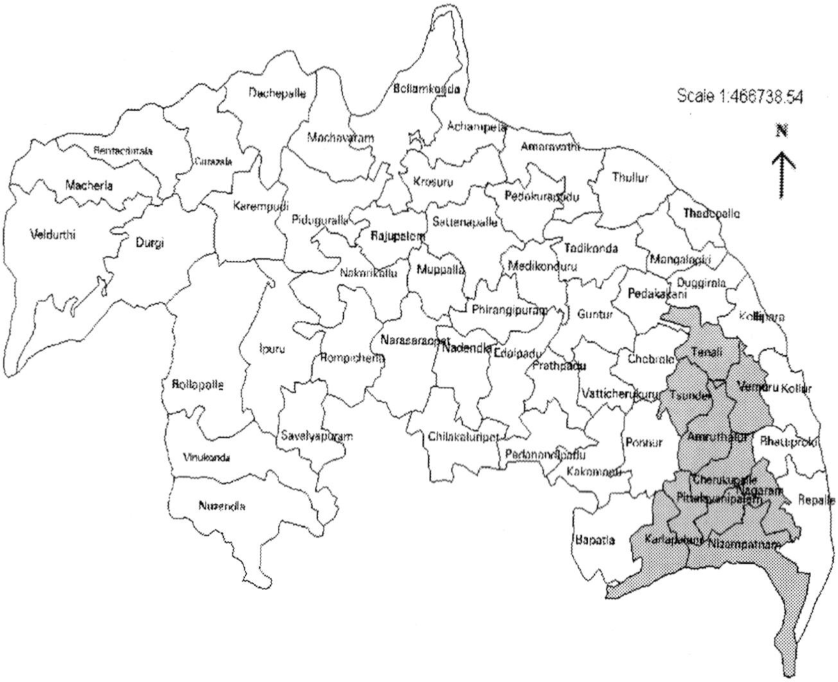

Fig. 2. The study area in the Guntur district considered for damage assessment due to Ogni storm

Remote sensing data

Pre event cloud free satellite image of IRS P6 LISS III was analyzed of November 14, 2005. The pre event and post event Envisat ASAR image of April 23, 2006 and November 4, 2006 were also analyzed. The Envisat ASAR images characteristics for each of these two dates is given in Table 1.

Table 1. Envisat ASAR image characteristics for each date

S. No.	Date of Envisat ASAR image	Polarization	Time	Beam modes (Ascending/ Descending)	Resolution (in meters)
1	April 23, 2006	HH, VV	10.30 am	Descending	50 m
2	November 4, 2006	VH, VV	10.30 pm	Ascending	50 m

Preprocessing of IRS P6 LISS III data

The IRS P6 LISS III image of November 14, 2005 was geometrically corrected. The exact area of nine mandals of Guntur district was extracted using a digitized standard mandal vector layer. For supervised classification, green, red and near

IR bands were considered *i.e.* Band 2, Band 3 and Band 4 of the IRS P6 satellite (standard FCC was generated). Using Supervised Classification and Maximum Likelihood technique, landcover classification was performed on IRS P6 LISS III image[16]. Six landcover classes namely, Water body, Paddy, Other crops, Current fallows, Aqua culture and Others were identified. Each pixel in the image was classified in any of these above six classes based on Supervised Classification and Maximum Likelihood technique.

The area covered by each of these classes is shown in Table 2. It can be seen from Table 2 that out of total of 115.989 thousand hectares area of nine mandals of Guntur district, 65.234 thousand hectares is covered by paddy crops. The landcover classified map of nine mandals of Guntur is given in Figure 3. The paddy area of nine mandals reported was 71 thousand hectares for the kharif season of 2006[14,15].

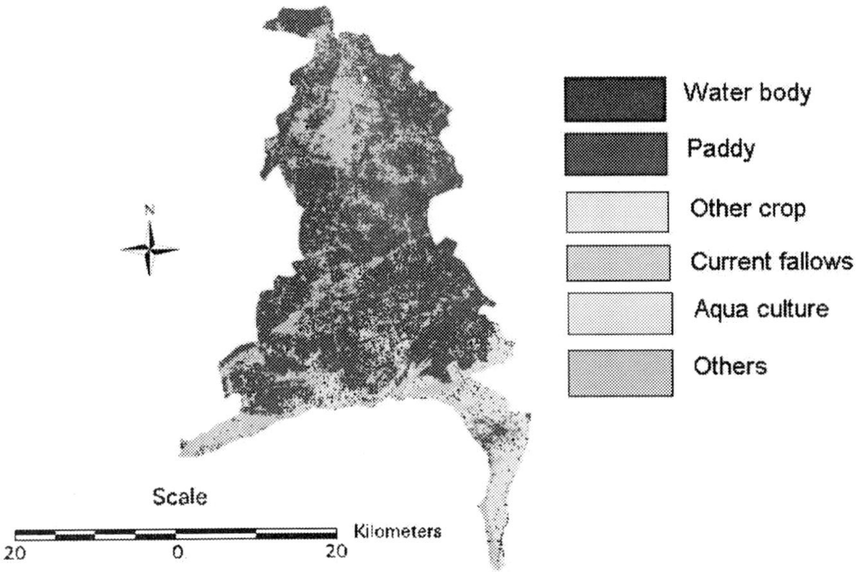

Fig. 3. Landcover map of nine mandals of Guntur district

Table 2. Land cover class and corresponding area of each class in nine mandals of Guntur district using IRS P6 LISS III classified image

Landcover class	Area in thousand hectares
Water body	1.362
Paddy	65.234
Other crops	5.087
Current fallows	30.944
Aqua culture	8.873
Others	4.487
Total area in thousand hectares =115.989	

Preprocessing of envisat ASAR data

The pre and post event images of Envisat ASAR were obtained from European Space Agency (ESA). The pre event image of April 23, 2006 was acquired in IS4 (swath), descending mode with HH and VV polarization, whereas post event image of Envisat ASAR of November 4, 2006 was acquired in IS4 (swath), ascending mode with VV and VH polarization. Both the scenes were obtained in Single Look Complex (SLC) format.

For comparison of temporal data for quantitative analysis absolute calibration is necessary. For DN to sigma-0 (it's nothing but backscattering coefficient) conversion, we have developed software using the algorithms[17]. Other processing steps such as slant to ground range and geocoding were performed using ENVI software.

As described in ESA document, the backscattering coefficient (σ^o) values are obtained by combining the radar pixel intensity and the incidence angle (I_j) at each pixel location, where the angle of incidence is obtained from the ASAR header data using interpolation method and is given in Equation 1.

$$\sigma_j^{\,o} = (DN^2/K) \times (R_d/R_{ref})^4 \times (1/G^2) \times \sin(I_j) \qquad (1)$$

where,

K is the absolute calibration constant,

DN^2 is the pixel intensity,

G^2 is two-way antenna gain at distributed target look angle,

R_d is distributed target slant range distance and R_{ref} is reference slant range distance.

The relationship between incidence angle (α_j), Earth angle (γ_j), slant range distance (R_j), satellite to Earth centre distance (R_{sat}) and elevation angle (θ_j) is given in Equation 2 and Equation 3 respectively.

The Earth angle (γ_j) can be derived for each range sample as:

$$\gamma_j = \text{asin}\{R_j(\sin(\alpha_j)/R_{sat}\} \qquad (2)$$

and the elevation angle θ_j is :

$$\theta_j = \alpha_j - \gamma_j \qquad (3)$$

where j increases from 1 to N.

All the images were multilooked in azimuth direction 10 times and the range direction 2 times to reduce the speckle noise. Later, the slant range is converted to ground range using ENVI software. The final spatial resolution of the data is about 20m × 20m with slight variation with incidence angle. Later, data were subjected speckle filtering using Enhanced Lee filter.

Finally, backscattering coefficient is expressed in decibel units by taking $10\text{Log}_{10}(\sigma^0{}_j)$. Only VV polarized images of April 23, 2006 and November 4, 2006 are considered for damage assessment analysis.

As we did not have topographic maps of study area for geocoding SAR images, we took four corner coordinates provided in the metadata of ENVISAT ASAR product and geocoded the images. Later image to image registration is done by identifying some GCPs in both the images. The range of incidence angle (degree) for April 23, 2006 and November 4, 2006 of Envisat ASAR images was 30.8-36.177 and 30.84-36.15 respectively.

After preprocessing of Envisat ASAR images, all the values were obtained in dB *i.e.* backscattering values. Then these images were imported to ERDAS Imagine (version 9.3) for further analysis. The geographical areas of nine mandals of Guntur district were extracted from April 23, 2006 and November 4, 2006 Envisat ASAR images using a standard digitized vector layer.

Deterministic model

Few pixels of pre event Envisat ASAR of April 23, 2006 (corresponding to November 14, 2005 IRS water classified pixels) were extracted and are given in Table 3. It can be seen from Table 3, the highest backscattering value obtained was –14.0 dB. Hence a threshold of –14.0 dB was selected to classify water and non-water in multi-temporal Envisat ASAR images. Backscattering values less than –14.0 dB were classified as water and greater than –14.0 dB were classified as non-water.

Table 3. The extracted water pixels backscattering values and count

Backscattering value	count
-17.3	1
-16.9	1
-16.5	1
-16.3	1
-16.2	2
-16.1	2
-16.0	1
-15.9	1
-15.8	1
-15.7	1
-15.6	7
-15.4	4
-15.2	1
-15.1	8
-15.0	4
-14.9	4
-14.8	2
-14.7	1
-14.5	3
-14.4	2
-14.0	1

Results and discussion

The water area in pre event Envisat ASAR image of April 23, 2006 and post event Envisat ASAR image of November 4, 2006 at threshold of –14.0 dB was 2.532 and 13.502 thousand hectares respectively.

Conclusions

The damage map generated using this tool is self-explanatory. Inaccurate landcover classification map (which forms the baseline for such study) using visible-IR bands would leads to errors in final submergence results. As discussed earlier, threshold of –14.0 dB was identified for classification of water and non-water area in pre event SAR using pre event classified IRS image. It was found that 80% of water pixels of pre event IRS and Radar classified images matched. Overlay of the cadastral map will help in exactly getting information of land parcels affected. This tool will find application for assessing damage due to other disasters like floods, Tsunami, Tornado *etc.*

Acknowledgement

Authours thank Human Dimension of Global Change, CMU, USA for sponsoring this research. Also, we thank Dr. Manab Chakraborty and N.K. Patel, Space Application Centre, ISRO, Ahmedabad for valuable suggestions and guidance. I thank my research colleagues Dr. Ajay Singh and Miss Upasna Sharma for constant encouragement.

References

1. Robinove C J, Interpretation of a Landsat image of an unusual flood phenomenon in Australia, *Remote Sensing of Environment*, 7 (1978) 219-225.
2. Wickware G M and Howarth P J, Change detection in the Peace-Athabasca Delta using digital Landsat data, *Remote Sensing of Environment*, 11 (1981) 9-25.
3. Chen P, Liew S C and Kwoh L K, Tsunami damage assessment using high resolution satellite imagery: A case study of Aceh, Indonesia, IEEE International Geoscience and Remote Sensing Symposium, 2 (2005) 1405-1408.
4. Kundu S N, Sahoo A K, Mohapatra S and Singh R P, Change analysis using IRS P4 OCM data after the Orissa supercyclone, *International Journal of Remote Sensing*, 22 (2001) 1383-1389.
5. Ormbsy J P, Blanchard J P and Blanchard A J, Detection of lowland flooding using active microwave systems, *Photogrammetric Engineering and Remote Sensing*, 51 (1985) 317-328.
6. Imhoff M L, Verimillion C, Story M, Choudhury A and Gafoor A, Monsoon flood boundary delineation and damage assessment using space borne imaging radar and Landsat data, *Photogrammetric Engineering and Remote Sensing*, 53 (1987) 405-413.
7. Schumann G, Hostache R, Puech C, Hoffmann L, Matgen P, Pappenberger F and Pfister L, High-resolution 3-D flood information from Radar imagery for flood hazard management, *IEEE Transactions on Geoscience and Remote Sensing*, 45 (2007) 1715-1725.

8. Jiren L and Shifeng H, Application of ERS/ENVISAT to flood monitoring and assessment in China, *IEEE International Geoscience and Remote Sensing Symposium,* 8 (2005) 5674-5677.

9. Yamada Y, Sakurai-Amano T and Takagi M, Detection of flood damaged areas in the entire Chao Phraya River Basin from JERS-1/SAR images with a help of spatial information, IEEE International Geoscience and Remote Sensing Symposium, 5 (2002) 2877-2879.

10. Badji M and Dautrebande S, Characterization of flood inundated areas and delineation of poor drainage soil using ERS-1 SAR imagery, *Hydrological Processes,* 11 (1997) 1441-1450.

11. Profeti G and Macintosh, H, Flood management through Landsat TM and ERS SAR data: a case study, *Hydrological Processes,* 11 (1997) 1397-1408.

12. Takeuchi S, Konishi T, Oguro Y, and Suza Y, (1999). Estimation of rice-planted areas in early stage using Radarsat data, http://www.gisdevelopment.net/aars/ acrs/ 1999/ ps2/ps20020a.shtml (last accessed on September 12, 2009)

13. Oguro Y, Suga Y, Takeuchi S, Ogawa M, Konishi T and Tsuchiya K, Comparison of SAR and Optical sensor data for monitoring of rice plant around Hiroshima, *Advance Space Research,* 28 (2001) 195-200.

14. Agriculture Production Plan, Agriculture Department, Guntur District, 2008-2009.

15. Director of Economics and Statistics, 2006, Government of India, Andhra Pradesh, Hyderabad.

16. Lillesand T M and Kiefer R W, *Remote Sensing and Image Interpretation,* (John Wiely and Sons, New York), Inc., 3rd Edition, 1994, pp. 750.

17. Rosich B and Meadows P, Absolute Calibration of ASAR Level 1 Products Generated with PF-ASAR, ENVICLVL- EOPG-TN-03-0010, Issue 1, date of issue: October 7, 2004, revision 5, Technical note http://earth.esa.int/pub/ESA_ DOC/ ENVISAT/ASAR/ASAR_products_absolute_calibration_v1.5.pdf, 2004 (last accessed November 15, 2009).

The increase in public awareness of climate change by use of ICT tools in Hadonahalli, Bangalore rural

T.R. Ramakrishnan[1] *& K. Puttaraju*[2]

[1]Department of Communication, [2]Faculty of Science, Bangalore University, Bangalore, India
[E-mail: [1]ramakrishnan@bub.ernet.in, [2]rama_krishna_tr@yahoo.co.in]

Introduction

Climate change is defined by United Nations Framework Convention on Climate Change (UNFCCC) as "A change of climate which is attributed directly or indirectly to human activity that alters the composition of the global atmosphere...". The elements in the definition include changes in daylight length, rainfall, weather, environment, and biodiversity that occur over a period of time. It also includes the influence of air and water pollution, influence of heat and cold in the ocean, greater intensity of storms and natural disasters, and human induced causes. It has great impact on agriculture, insect and pest prevalence, rainfall patterns (water cycle), crop failure, drying up of water sources, natural disasters, availability of food and many more. The impact on poor countries will be greater than rich countries.[1]

Peter Newell (2010)[2] brings to light that Climate change first was registered on the radar screen of fossil fuel companies in the late 1980s and early 1990s amid growing demands for an international treaty to address the issue. The UN's Intergovernmental Panel on Climate Change (IPCC) had been set up in 1988. Years had passed with keen watch on climate change from then to the present day where tomorrow's generations have been made aware about the dangers of environment disasters.

Peter notes B Parks[3] statement that National disasters become social disasters because poorer people tend to live in flimsier housing, nearer more polluting industry and in neighbourhoods with inadequate infrastructure. Climate change reproduces this tendency for environmental change to magnify existing social inequalities. Hence, the climate justice movement has evolved and is progressing.

Nevertheless, there has been tremendous progress made by mega-organizations. At the UN Millennium Summit in September 2000, the eight Millennium Development Goals (MDGs)[4] set worldwide objectives for reducing poverty and hunger, improving health and education, empowering women and ensuring environmental sustainability by 2015. The UN is convening a special summit in New York, 20-22 September 2010, to agree on a plan to accelerate global action on the Goals.

World Meteorological Organization (WMO) mention that the application of climate knowledge in risk management strategies helps alleviate poverty and support human welfare and safety. Climate, the environment and sustainable development are inextricably linked. MDGs Goal 7[5] declare that a nation should ensure environmental sustainability by integrating the principles of sustainable development into country policies and programmes. It emphasizes on reverse loss of environmental resources. Further, it targets to reduce by half, the proportion of people without sustainable access to safe drinking water. When the results were viewed it was reported that under MDG's, Goal 7, the UN reports that over the last 10 years the world lost 13 million hectares of forest each year. The population growth have produced a nearly 50 per cent increase in global CO_2 emissions between 1991 and 2007, from 21.9 to 29.6 million metric tons. The report interpret that unless decisive action is taken, emissions will again raise rapidly as the world economy reboots from its slumber during in 2008-10.[6]

The World Climate Conference-3 (WCC-3), held from 31 August to 4 September 2009 in Geneva, decided to establish a Global Framework for Climate Services (GFCS) to strengthen the production, availability, delivery and application of science-based climate monitoring and prediction services. GFCS is designed to mainstream climate science into decision making at all levels and help ensure that every country and every climate-sensitive sector of society is well equipped to access and apply the relevant climate information.

Based on internationally agreed development goals, including those in the Millennium Declaration, which are premised on international cooperation, indicative targets may serve as global references for improving connectivity and access in the use of ICTs in promoting the objectives of the Plan of Action, to be achieved by 2015. One among the strategies were to connect villages with ICTs and establish community access points.[7].

A Global humanitarian forum report estimate that climate change is responsible for some 300,000 deaths each year and over USD 100 billion worth of economic losses due to shocks to health and agricultural productivity. Michel Jarraud, Secretary-General of the WMO, the United Nations System authoritative voice on Weather, Climate and Water, which is coordinating involvement of National Meteorological and Hydrological Service (NMHS) participating in the initiative, Mr. Jarraud said that, "for food production, almost every decision is linked to weather, climate and water parameters. Working through NMHSs, WMO will identify weather information needs, advise on technical requirements and help disseminate the information. This initiative would help create better weather information."[8]

Jayalakshmi Chittoor (2008)[9] mention that it is evident that the Intergovernmental Panel on Climate Change (IPCC) has made more public awareness on Climate change.

Dr. Shahriar Hossain (2009)[10], finds that adaptation to climate change is a key necessity for the global community. To adapt using ICTs for improvement of climate change can be done with the involvement of civil societies, local communities along with government agencies. Shahriar concludes stating that public awareness, education and training will be most critical tool to involve all sections of the public in the process.

Gokul Raj M.P (2005)[11] finds on the study of the role of radio and newspaper in dissemination of agricultural technology in Karnataka that majority of the radio listening farmers belong to the category of young age, low education, medium family, low cosmopoliteness, high extension contact, medium attitude, low credibility, low mass media participation, and high sources of information use besides listening to Krishiloka programme of radio. With respect to newspaper readers it was found that most of them were found in the old age, high education, medium farm and family, high cosmopoliteness, high attitude and high social participation, extension contact, low credibility, high mass media participation and high sources of information and besides listening to Krisholoka programme over radio. It was also suggested that studies on emerging information technologies could be explored in rural Karnataka.

The above referrals from large scientific bodies on facts associated with climate change just necessitate the awareness study on climate change on use of ICT tools. Hence, a village in Karnataka namely Hadonahalli at Bangalore Rural was selected. Hadonahalli is a select locale of the University of Agricultural Sciences, Bangalore for all its agricultural extension activities on scientific farming among farmers. Hadonahalli is located close to the University. Hence, there are probabilities that the farmers could be considered as advanced farmers.

Problem

The Indian farmer till date is believed to follow the traditional Indian calendar for weather forecast. The farmer considers the scientific weather forecast information given in the newspaper, radio and Television as a compliment and sometimes as warnings. The farmer would discuss about rainfall during the previous years from farming friends and locals and conclude to look up in haste to the sky and lay trust in God for the rains to pour.

Hence, to take strategic decisions during the stages of farming, timely scientific weather forecast is necessary for the farmers. The farmers are becoming aware about the global climate changes and its impact on the stages of farming. The strategic indecisions cost the farmer revenue losses, which in many cases have led them to take huge loans that are unpaid. Hence, some have committed suicide others have fled from farming to various professions. Nonetheless, some continue to farm by meeting the expenses from their migrated earnings others sell for real estate development or lease it out.

It is assumed that farmers seldom follow the scientific weather forecast. It is time they adapt to scientific weather forecast to take timely farming decisions. Hence, the present study focuses on the public awareness of climate change by use of ICT tools in Hadonahalli Village, Bangalore Rural.

Objectives

1. To identify the mega-bodies on climate change and its role.
2. To study the process of weather forecast from the Government in Bangalore.
3. To analyse the ICT initiatives of the Department of Agriculture, Bangalore.
4. To measure the awareness level of scientific weather forecast using ICT tools.
5. To suggest a new model on dissemination of communication of weather forecast.

Methodology

The methodology used in conducting the study has been as follows. Secondary data was referred from select websites, books and journals to understand the first three objectives. The practices of weather forecast and rainfall measurement was understood from discussions held at Indian Meteorological Department (IMD), Bangalore and Karnataka State Natural Disaster Monitoring Centre (KSNDMC), Bangalore. Primary data was collected using a questionnaire that was administe-red among the public, predominantly farmers and farm workers at Hadonahalli Hobli in Tubugere Taluk, Bangalore rural in order to understand the awareness level of the climate change upon using Information and Commu-nication Tech-nology. This hobli was selected as the ICT initiatives of University of Agricul-tural Sciences is exercised at the Village Resource Center situated at Hadonahalli.

A survey method based on a questionnaire reflecting to the objectives of the study was administered. The sampling method adopted in the selection of the respondents is simple random sampling.

The increase in public awareness towards climate change and use of ICT tools were the two dependent variables. Independent variables being crop, present decisions of weather forecast for farm activities, requirement of weather forecast at different stages of cultivation, decisions on scientific weather forecast, reasons for climate change, source of information and frequency of use of ICT tools. Empirical measurements (scales) were used in the study.

Data analysis was done using percentage. Percentage were worked out and used to make comparisons in the selected variables among farmer's practice of weather forecast, need and use of scientific weather forecast, the awareness of climate change and the use of ICT for accessing information about weather forecast and climate change.

There are three main sources for weather forecast namely (a) Indian Meteorological Department (IMD), (b) Rastriya Panchang and Krishi Panchang, (c) Drought Monitoring Cell (DMC).

Mega-bodies

The mega-bodies such as UNFCCC, IPCC, WMO, WCC-3, New Economics Foundation, UN Millennium Summit and many others whose roles have been referred in the background facts of this study have created a general awareness of climate change among the public in a short span of time. In Karnataka, there are two bodies that monitor the weather and transmit the data on a day to day basis to Geneva.

Indian meteorological department (IMD)

Meteorological Center, Bangalore, undertakes real time climate monitoring and publication of climate diagnostics bulletins and reporting of major anomalous climate events. Special climatological studies unit prepares the climatological summaries for user's interest and brings in the form of publications useful for different users who need climatological information.

Rainfall climatology for agricultural planning unit calculates appropriate sowing dates over different regions for the optimum crop production and brings in the form of a publication. This information is given to the State Government for its proper dissemination to the farmer community so that they may judiciously select proper sowing dates for a particular crop for the better agricultural production. Similarly IMD prepares biweekly aridity anomaly reports and maps for all seasons. Further, quantitative crop yield forecast and observations on the draught scenario are published.[12] The media available for the transmission of bulletins to the public are electronic (radio, TV) for communication of the products, Agrimet division's website for displaying Agromet Advisory Services (AAS) products in local language and Tarang service of Bharat Sanchar Nigam Limited (BSNL) for obtaining taluka level information.

Meteorological center at Bangalore has 300 centers situated at various State Government offices, schools across Karnataka. Mrs.Bharathi Gopalan of IMD mentions that they are manned by local men who generally have less education. They collect the rainfall data in physical mode measuring in millimeters (mm) using Ordinary Rain Guage (ORG) daily at 8.30 AM. This data is recorded in cards and ledger by the data collector at the centers and informed by phone to the meteorological center at Bangalore. The all weather data (sunshine, rain, clouds, wind, temperature and humidity) is entered in software namely M.K Software, on analysis the district level basis MIS report is extracted daily and filed. The report has all weather information (sunshine, rain, clouds, wind, temperature and humidity) and comparative analysis information. This information is available for distribution. The information is disseminated within the department using Virtual Private Network (VPN) terminals. It is sent to the regional office at Chennai, headquarters at New Delhi and to Geneva. The information is uploaded to the IMD Bangalore web site (Table 1).

Table 1. The web site imdbangalore.gov.in provides the following information

Name of report	Detailed report
Karnataka weather on local forecast	
Daily weather report	Summary and forecast, observational data, local forecast
Weather warnings	Port, Cyclone, 2ndStage, Fisheries, temperature, heavy rainfalls, costal bulletin
Weekly weather report	Weekly summary,Weekly statistics
Agri advisory bulletin	In Kannada and English, weekly map and seasonal map
Sunrise and sunset tables	Gadag, Bangalore, Mysore, Mangalore, Gulbarga, Devanahalli
Rainfall statistics	Winter(Jan-Feb), Pre-Monsoon(Mar-May), Monsoon (Jun-Sep), Post Monsson(Oct-Dec)
Rainfall time series(Graph)	Karnataka, Coastal Karnataka, North Interior Karnataka, South Interior Karnataka
Observations and Forecast	AWS data, Observational data, Chief Synoptic features, Weather warnings, Forecast for the next 5 days, Local forecast for all cities, Daily weather report, Weekly weather summary, weekly statistics/map, Seasonal statistics/map, Sunrise and sun set for main cities
Agro meteorology	District level forecast, District agro met advisory
Weather warnings	Cyclone page, Severe weather warnings
Seismology	Latest earthquake report, information on seismology
Airport information	BAIL, Bangalore, HAL, Bangalore, Belgaum, Hubli,Mangalore

Source: www. imdbangalore.gov.in

In addition, it has information on press releases, numerical models, satellite images, Doppler Radar products, Weather charts and models. The Daily summary of observations recorder consisting of details related to earthquake, temperature, humidity, rainfall and weather remarks in all districts of Karnataka as well as New Delhi, Kolkata, Mumbai, Chennai, Pune, Hyderabad and Thiruvanthapuram. Dr. Ramdas of IMD Bangalore reveal that this information is daily emailed to the following departments in Karnataka: Chief Minister's Office, all Government control line departments, Department of Agriculture, District commissioner, Tahsildar, Agricultural officers, Assistant Director of Agriculture, Joint Director of Agriculture, Governor, Home Secretary, Secretary–Relief (disaster management), Secretary-Revenue, Commissioner-Bruhat Bengaluru Mahanagara Palike (BBMP), State intelligence, Nandi Infrastructure Corridor Enterprises (NICE), Media: Doordarshan and all private regional T.V. channels, Mysore Printers Pvt. Ltd., Indian Express, The Hindu, Smukthakarnataka, Udayavani, Kannadaprabha, Vijayakarnataka, UNI, AIR and other radio channels, Agro advisory unit of each district under Agrismet, IMD Chennai and the website. Totally about 40 departments are informed from IMD everyday about the observations recorded. The farmers can get the information by calling 1800-180-1717 or 080-22211118 or Agro Advisory Bulletin twice a week. The same details are also available on the IMD, Bangalore website which is constantly updated. The department receives on an average about 5 telephone enquiries from farmers which needs to be enhanced by creating awareness. Mrs.Bharathi states that there should be a common minimum programme for awareness and the human element should not be removed for monitoring and measuring weather data. The reason being due to extreme weather conditions the automated equipments could malfunction and it would take a while before they are rectified.

The second source for weather report information is the Rashtriya panchang and Krishi Panchang.

Rashtriya panchang

The Rashtriya Panchang contains astronomical and panchang information on sidereal time, equation of time, timings of sunrise, sunset, moonrise, moonset, tithi, nakshatra, yoga, karana, transit time of the sun, moon and planets to nirayana rasis and nakshatras, beginning of lagnas for central station of India as well as nirayana longitudes of the sun, moon, plantes and rahu. All of these have been calculated on modern scientific basis.

The symbols • ☽ ○ ☾ indicate new moon, first quarter, full moon and last quarter respectively. Majority of the farmers follow the national calendar. All the regional language calendar provide a brief of the panchang. It is assumed that the farmers follow the panchang for keeping track of the weather forecast as per their experience and perceptions. [13]

Krishi panchang

The contents of the proposed Krishi-Panchang in Kannada has similar information of Rashtriya panchang which changes yearly includes the daily/monthly/annual weather forecasting for the particular year, crop prospects of that year based on planetary positions, season-wise crop strategy based on anticipated weather. Auspicious moments for agricultural and allied activities and some general agricultural guidance are included.

The third source is the rainfall measurement by the Drought Monitoring Cell.

Karnataka state natural disaster monitoring cell (KSNDMC)

The KSNDMC[14] was established in 1988 and it provides scientific input towards the reactive and proactive draught management programmes in the State. Data on drought indicators are collected, updated and analyzed for preparation of weekly report on rainfall pattern, rainfall deficiency for agriculture requirement, dry spells, agricultural operation and crop condition, moisture stress, minor irrigation tank status and reservoir levels in the state. The information in the form of reports is disseminated to line departments up to taluk level and other institutions including media. The Drought Monitoring Cell (DMC) of the Karnataka State Natural Disaster Monitoring Centre (KSNDMC), a registered society of the State Government, is on the verge of completing the ambitious project of installing telemetric rain gauges in hobli headquarters. Rain gauges have been installed in 607 hoblis and they will be installed in the remaining 140 hoblis in two months.[15] The DMC is implementing the project under which telemetric rain gauges will be installed in 4,500 gram panchayats with funding from Rashtriya Krishi Vikasa Yojane, a Centrally-assisted scheme. Once the installation of gauges is completed, Karnataka will become one of the few States in the country to implement the state-of-the-art system of measuring rainfall on real time basis.

The web site, as in Table 2, of DMC is updated periodically and ensured disseminations of information generated to all the user agencies. The three sources are observant and continually disseminate scientific weather forecast to the public.

Table 2. The website www.dmc.kar.nic.in provides information on the following

Activity	Type of Information
Rainfall and Agricultural conditions in the state	Weekly weather and crop report, Monthly Rainfall and agricultural situation, Seasonal rainfall and agricultural situation, annual rainfall and agricultural situation,
Geographical Presentation	Weekly rainfall pattern, monthly rainfall pattern, seasonal rainfall pattern, annual rainfall pattern, Rainfall map: State map with district boundaries, and District map with taluka boundaries, State map with taluka boundaries.
Query analysis	Rainfall data analysis, agricultural data and reserviour data
KSNDMC Highlights	Daily rainfall and forecast map, Aridity index and anomaly, Moisture adequacy index, weekly highlights and maps
KSNDMC reports	Notifications, Study report, Draught affected talukas, Satellite cloud images
Draught scenes	Draught affected scenes, Draught proofed scenes

Source: www.dmc.kar.nic.in

Beyond the above, it has links to Storage tanks in state , Daily weather report, Earthquake report, District-wise Rainfall maps, Taluk level rainfall pattern as on yesterday and District-wise moisture adequacy index.

The present study investigates the need for scientific weather forecast at Bangalore Rural.

Addressing the International Telecommunication Union (ITU) Telecom World 2009, Geneva, UN Secretary-General Ban Ki-moon[16] stressed the importance of information and communication technologies (ICTs) as part of the solution to the threats imposed by climate change not only to cut emissions and help countries adapt to the effects of climate change, but also to raise awareness of climate change. Secretary-General Ban highlighted specific examples of the use of ICTs, such as the use of mobile phone technology as a tool to help farmers be better informed about weather conditions. He said this project involves a partnership between the UN, mobile phone companies and other partners, through which 5,000 new weather stations were installed across Africa to monitor the impact of climate change and text messages are sent to farmers' mobile phones to provide them with relevant information.

Bangalore rural

Bangalore Rural District is located in the South-Eastern corner of Karnataka State and has 4 taluk: Devanahalli, Doddaballapura, Hosakote and Nelamangala. Doddaballapura has 5 hoblies namely Doddabelavangala, Thubagere, Sasalu, Madure and Kasaba. It has 29 grama panchayaths as given in Table 3.

Table 3. Agro climatic zone of Bangalore rural

Name of the zone	Districts (no. of Taluks)	Rainfall range in mm	Elevation in meters	Soil
Eastern Dry Zone (24 taluks)	Bangalore(R) (4), Bangalore U (3), Kolar (11), Tumkur(2)	670 to 888.9	800-900 in major areas, in remaining areas 900-1500	Read loamy soils in major areas, clay lateritc sils in remaining areas

Source: Agri Doctors Diary 2010, Karnataka Agriculture Graduate Officers Association

The Thubagere hobli is located in region IV (Southern Dry Region) and zone V (Eastern Dry Zone) according to agroclimatic classification. It lies between 77°34´ E longitude and 13°22´ N latitude at a height 1023 meters from mean Sea Level.

There are about nine schools in Hadonahalli with dry land of 1310 ha and irrigated area of 100Ha and forest area of 95 ha in 11 villages covering 2076Ha of 1267Ha cultivated area with 1131 families residing at Hadonahalli as given in Table 4. Table 5 provides the various crops grown in Hadonahalli during the two seasons South West Monsoon (Mungaru Malai) and North Eastern Monsoon (Hingaru Malai).

Table 4. Panchayat wise villages, geographical area, cultivated area and number of farm families in Tubagere Hobli

S. No	Village panchayat	Villages	Geographical area (Ha)	Cultivated area (Ha)	No. of families
1	Hegdehalli	14	1762	1324	1127
2	Melekote	15	2175	1504	1891
3	Hadonahalli	11	2076	1267	1131
4	S.S.Ghatti	12	4054	2535	1516
5	Tubagere	20	3895	8119	1835
	Total	**72**	**13990**	**9469**	**7500**

Source: Baseline report (2006-07) DBT-Rural Bio-Resource Complex Project, UAS, Bangalore

Table 5. Area under different crops(Ha) in Hadonahalli

Crop	Ha
Ragi	700
Paddy	40
Maize	167
Ground nut	50
Sunflower	20
Cow pea	10
Redgram	46
Field bean	40
Green gram	1
Black gram	1
Soyabean	1
Horsegram	110
Niger	45
Mustard	23
Sesamum	1
Vegetables	25
Mulberry	28

Source: Baseline report (2006-07) DBT-Rural Bio-Resource Complex Project, UAS, Bangalore

From observations made at the computer department of the Head office of the Department of Agriculture, Government of Karnataka, it is very clear that the Agrisnet system is progressing well with almost all the districts entering and updating the data under respective modules. However, the farmers do not have access to any of the details independently. The 12 modules under Agrisnet are installed and operational at the Taluk level in Karnataka.

Some of the problems faced during the execution of the projects were on data processing processes. There is also lack of manpower and computer hardware at Village level and Taluk level for data entry both in Kannada and English.

The Nemmadi Project was set up by the Government of Karnataka through a Public Private Partnership (PPP) model, a network of 800 telecenters at village Hobli level. It has set up 177 back offices at taluka level. These telecenters deliver range of G2C and B2C services.

Nevertheless, it is the Agrisnet that has great impact among G2G communication at the Department of Agriculture, Government of Karnataka. The present communication model monitored at the department head quarters is displayed below.

Present communication model for agrisnet project

State Government of Karnataka

↕ {There is two way communication using ICT tools}

Department of Agriculture

↕ {There is two way communication using ICT tools}

Taluk Level Department of Agriculture

↕ {Only one way communication using ICT tools}

Hobli Level Department of Agriculture

↕ {Only Verbal or physical paper based inter-personal communication}

Farmers

It is found there is a digital divide between Hobli level department and Taluk level department and between farmers and Hobli level department communication. To bridge the gap there are many pilot initiatives taken by University of Agricultural Sciences, Krishi Vigyan Kendras, Mass Media and Raita Samparka Kendras to reach out to the grassroot level. [17]

Ramakrishnan, T.R. (2009)[18], conclude that ICT based communication by few experts to many farmers at Village Resource Centers (VRCs) require proper communication model for confidence building like the one suggested as inclusive public relation model of communication. The factors that motivate the farmers are the ambience at VRC, a meeting point for farmers, refreshments, friendly attitude of the staff, proximity to the center, signing the attendance register, scheduling material distribution for farming and collecting subsidies.

The farm advisory services from Agricultural Technology Information Center (ATIC) at UAS, GKVK campus had provided 4823 advisory services though 194 personnel letters, 1478 telephone calls, 949 office calls, 348 consultations, attended 1778 general visitors besides 344 calls of Kissan Call center. [19]

The Foundation for Agrarian Studies (FAS)[20] had done a survey between 2005 and 2009 and their findings were that ownership of TV dominated other durables and services with 50% ownership in most villages in India. In two villages in Karnataka Alabaujanahalli (Mandya District) and Siresandra (in Kolar District) where the incidence of TV ownership was generally high. Telecom Regulatory Authority of India (TRAI) show that teledensity in rural India was 24.29 in March 2010 compared to 119.73 for urban areas. The findings of the study reveal that there is a clear indication that general economic "prosperity" is not good enough to even out extreme inequalities.

The Hindu[21] reports that Smart farmers in Tubagere hobli of Doddaballapur taluk use the hi-tech Automatic Weather station installed in their area beyond agricultural operations. Several smart farmers have started referring to the weather forecast for non-agricultural events such as marriage etc. Dr.Narayana Gowda vice chancellor of UAS, states that the automatic weather forecast data is analyzed by weather and agricultural experts of UAS-B and passed on to farmers through the 49 village-level milk co-operative societies in the hobli who then pass the word around about the forecast. However, the data is available at the local Krishi Vijnana Kendra.

The difference shown in the Table 6 is due to the non collection of data on holidays and days taken off by the manual rain collector. However, there are also reasons that the automated telemetric gauge equipment was malfunctioning. Nevertheless, the system is trying its best to achieve accurate measurement without discrepancies between both the systems.

Deccan Herald[22] reported that Panchatantra, a new web based system enables gram panchayaths in Karnataka to host on the web all the information they have in records by logging on to http://stg1.kar.nic.in/panchatantra. As many as 5,220 of the 5,626 gram panchayaths in the state have gone online.

A Vaidyanathan (2010)[23], mentions that researchers outside the National Agricultural System (NAS) and those within the system did not consider technology important to undertake an objective analysis of technology and avoided writing about the downside of technology for fear of displeasing agro-biologists, who dominated the system and also happened to be their leaders or bosses. The system did not encourage sharing of information for critical assessment of technology.

Table 6. Rainfall measurement using automatic telemetric rain gauge and Ordinary rain gauge in Doddaballapura Hobli, Bangalore Rural between June 2010 to August 2010 which covers the neighboring village Hadonahalli.

Month	Method	Cumulative(mm)	Difference(mm)
June	TRG	235.5	
	ORG	228.8	6.7
July	TRG	128	
	ORG	124.3	3.7
August	TRG	152	
	ORG	151.5	0.5

Source: The Drought Monitoring Cell (DMC) of the Karnataka State Natural Disaster Monitoring Centre (KSNDMC)

There are several studies highlighting the use of ICT for the upliftment of the common man. This study has looked into the awareness of climate change especially focusing on the shift of trust from perceived notion on weather forecast from past experience and the Rashtriya panchang to scientific weather forecast.

Findings about the awareness study

The study on awareness about climate change was conducted using survey methodology among Ragi growers of Hadonahalli hobli at Bangalore Rural. The findings are elaborated in Tables 8 to 19.

Table 7 project that 57% of the farmers predict the rainfall from their past experience and 20% from their self imagination. It is found that only 13% follow the calendar and 10% indulge in conversation with other farmers to take a decision on farming process based on rains.

Table 7. Opinion followed for weather forecast

From general conversation	10%
From past experience	57%
From the Calendar	13%
From self imagination	20%

The Table 8 shows that 81% of the respondents are aware that it would rain when they observe the small and large ring around the moon stating that there is a depression in the ocean. They also know that when there is excess heat on an ordinary day, it may rain. However, they are not at all aware about how pocket area rainfall happen.

Data from Table 9 discloses that 100% of the farmers give extremely high importance on rainfall forecast at all the stages of farming. 88% of the respondents have not at all given importance to Wind forecast in almost all stages except during harvest, and 63% during disease management time and 38% during pest management season.

Table 8. Awareness of rainfall based on situations

Depression (Ocean)		**Ordinary day (due to heat)**		**Pocket area (Certain areas)**	
Y	**N**	**Y**	**N**	**Y**	**N**
81%	13%	66%	34%	0%	100%

The data shows that temperature forecast is extremely important during preparing the land for sowing and harvesting as 100% responded to the two variables. The table display that 13% gave somewhat importance during pest management and 34% during disease management. It can be concluded that rainfall forecast is the most important factor and the farmers need to be informed regularly about rainfall.

Table 10 disclose that almost 72% of the respondents require rainfall forecast within 24 hours, 25% stated that within 48 hours is fine and 3% did not mind being informed 3 days in advance. None of the respondents wanted 5 days advance weather forecast.

Table 9. Importance given to weather forecast for different stages of farming. n=32

Stages of farming	Rainfall					Wind					Temperature				
Not at all (1), Not very (2), No opinion either way (3), some-what (4), extremely (5)	5	4	3	2	1	5	4	3	2	1	5	4	3	2	1
Preparing the land for sowing	32	0	0	0	0	0	0	0	0	32	32	0	0	0	0
Spacing & Sowing	32	0	0	0	0	0	0	0	0	32	0	0	0	0	32
Application of manure and fertilizers	32	0	0	0	0	0	0	0	0	32	0	0	0	0	32
Weed management	32	0	0	0	0	0	0	0	0	32	0	0	0	0	32
Water management	32	0	0	0	0	0	0	0	0	32	0	0	0	0	32
Pest management	32	0	0	0	0	12	20	0	0	0	0	4	0	0	28
Disease management	32	0	0	0	0	20	12	0	0	0	0	11	0	0	21
Harvesting	32	0	0	0	0	28	4	0	0	0	32	0	0	0	0

Table 10. Period of rainfall forecast expected by farmers

24 hours	48 hours	3 days	5 days
72%	25%	3%	0%

The Table 11 illustrate that predictions based on farmers intuitions does not work and 78% of the respondents vouched for it and about 22% mentioned their predictions based on their intuitions were always right.

Table 11. Decisions being right based on intuition.

Always	Sometimes	Never
22%	78%	0%

Table 12 point out that 88% farmer's decisions were right for the South West monsoon season (Mungaru Malai). But only 13% of their predictions were right during North-Eastern monsoon (Hingaru Malai).

Table 12. The time that farmers predictions were right

Pre-monsoon	South-west monsoon	North-East monsoon
88%	12%	0%

Table 13 expose that all the farmers need proper rainfall forecast during all the stages of farming. It shows that 78% of the respondents give extreme importance for wind forecast during harvest season and 56% during disease management. However, 84% mentioned that temperature forecast is extremely important during harvest but 63% had no opinion if they need temperature forecast during disease management.

Table 13. Importance given for the need of proper weather forecast at different stages of farming.

Stages of farming	Rainfall					Wind					Temperature				
Not at all (1), Not very (2), No opinion either way (3), some-what (4), extremely (5)	5	4	3	2	1	5	4	3	2	1	5	4	3	2	1
Preparing the land for sowing	32		0	0	0	0	0	0	0	32	0	0	0	0	32
Spacing & Sowing	32		0	0	0	0	0	0	0	32	0	0	0	0	32
Application of manure and fertilizers	32		0	0	0	0	0	0	0	32	0	0	0	0	32
Weed management	32		0	0	0	0	0	0	0	32	0	0	0	0	32
Water management	32		0	0	0	0	0	0	0	32	0	0	0	0	32
Pest management	32		0	0	0	0	0	0	0	0	0	22	0	0	09
Disease management	32		0	0	0	18	4	5	2	0	0	0	20	0	12
Harvesting	32		0	0	0	25	5	2	0	0	27	5	0	0	0

It can be concluded that rainfall forecast must be given at hobli (village) level. The present system follow forecast at district level. The wind forecast maybe required only when there is extreme climate change. The temperature forecast should compliment the daily rainfall forecast.

It was good to know that all the respondents followed the mass media to be informed about the scientific weather forecast. But only few followed the scientific forecast seriously for their agricultural decisions as the mass media reports only district wise forecast which seldom has an effect at hobli.

Table 14. Mass media as a source for weather forecast.

Newspaper	Radio	T.V	Mobile phone	Internet	Campaigns	Phone	Government office (VSAT)
56%	31%	100%	0%	0%	0%	0%	63%

The above table 14 disclose that all (100%) of the farmers have access to Television viewing and 63% to the Krishi Vighyan Kendra office where the scientific forecast is received by email, fax, telephone or VSAT from IMD.

Though 56% follow the newspaper and 31% listen to the radio, these two medias have not been taken too seriously by the farmers when compared to T.V or the VSAT programme at the KVK. It can be concluded that the 2 vibrant mediums be strengthened by providing hobli level forecast more regularly through mass media.

All of the respondents followed the scientific weather forecast during North East monsoon (Hingaru Malai) and 59% follow during South West (Mugaru malai) from the mass media.

Table 15. Frequency of taking decision at different farming stages based on the scientific weather forecast from mass media

	Rainfall			Wind			Temperature		
Stages of farming	A %	S %	N %	A %	S %	N %	A %	S %	N %
Preparing the land for sowing	69%	31%	0%	0%	0%	100%	0%	0%	100%
Spacing & Sowing	0%	41%	59%	0%	0%	100%	0%	0%	100%
Application of manure and fertilizers	88%	13%	0%	0%	0%	100%	0%	0%	100%
Weed management	0%	0%	100%	0%	0%	100%	0%	0%	100%
Water management	78%	22%	0%	0%	0%	100%	0%	0%	100%
Pest management	0%	13%	88%	0%	0%	100%	0%	47%	53%
Disease management	0%	0%	100%	0%	0%	100%	0%	0%	100%
Harvesting	38%	44%	19%	0%	75%	25%	0%	0%	100%

Always(A) Sometimes (S) Never (N)

The Table 15 project that 69% of respondents always followed the scientific rainfall forecast during preparing the land for sowing, 88% during application of manure and fertilizers, 78% during water management and 38% during harvesting. 31% sometimes followed during preparing the land for sowing, 41% during spacing and sowing, 13% during application of manure and fertilizers, 22% during water management, 13% during pest management and 44% during harvesting. All the respondents mentioned that during disease management and weed management they never followed the rainfall forecast. The wind forecast was somewhat important for them to decide during harvest but only 25% were not keen about it. Only 47% were somewhat keen on the temperature forecast during pest management stage. But none of them were keen to have temperature forecast for any of the other stages of agriculture.

It clearly indicates that majority follow the scientific weather forecast on rainfall from mass media during preparing the land for sowing, application of manure and fertilizers, water management and harvesting. Majority do not follow the wind and temperature forecast from the mass media. The reason for them not to follow are that it does not relate to their present traditional methods of farming. Scientific reasons such as pest management related wind and grains management related to temperature have not reasoned to their current practice which need to be informed to the farmers.

Table 16 projects that only 66% sometimes trust the scientific weather forecast, while 34% always believe the scientific forecast.

Table 16. Percentage of farmers who trust the scientific weather forecast.

Always	Sometimes	Never
34%	66%	0%

It was found that all the respondents were aware about climate change. And 88% of them felt the change in climate in their practice in agriculture. Hence, all who had an experience on climate change, feel that they should start believing the scientific weather forecast in future than their present method of following the weather forecast for decision making on farming process at different stages of agriculture.

Table 17 reveal the overall awareness of climate change phenomena. There is a higher percentage (88%) of farmers who are aware about change in atmosphere, 16% on change in hydrosphere, 9% about external changes, 12% on changes in solid earth, 100% related to rises in sea levels, 6% are aware about change in plant and animal existence, no one is aware about change in land-sea boarders or change from treeless area to temperate forest or change from temperate forest to deserts, or change in extreme temperature. However, all are aware about change in rainfall pattern. And none are aware about change in wind. About 16% were aware about change in draught, 6% in change in human population and 81% in change in irrigation water supply or ground water supplies.

Table 17. Awareness of climate change phenomena'sn=32

Climate change phenomena	% Aware	% Not aware
Change in Atmosphere (Air, water vapour, CO_2, clouds, particles)	88%	13%
Change in the Hydrosphere (continental ice-sheets, mountain glaciers, seas ice and surface snow cover)	16%	84%
External changes (Volcanic eruptions, Sun's output, Earth's orbital variations)	9%	91%
Changes in solid earth (continental drift, mountain building, erosion and siltation)	12%	88%
Large rises in sea levels	100%	0%
Massive changes in plant existence	6%	94%
Massive change in animal numbers	6%	94%
Massive change in Land-sea boarders	0%	100%
Change from treeless area to temperate forest	0%	100%
Change from temperate forest to deserts	0%	100%
Change in temperature (Extreme temperature/climate)	0%	100%
Change in rainfall (Floods, less rainfall during rainy seasons)	100%	0%
Change in wind	0%	100%
Change in Draughts	16%	84%
Change in human population	6%	94%
Change in irrigation water supply	81%	19%

It can be concluded that the response noticed that majority of farmers are aware about climate change caused due to change in atmosphere, changes in solid earth, large rises in sea levels, change in rainfall and change in ground water supply. But they are not aware of climate change caused due to the other factors mentioned in Table 18.

The ground water in surrounding areas such as the textile hub of Doddabalapura is polluting the soil and ground water for proper usage of soil for cultivation. This is causing danger to the once fertile soil used for grape farming in the same taluk.

Table 18 show that 100% of the respondents view Television programmes and watch the weather report. About 47% knew about it from listening to radio programmes and 38% from various Government led programmes in Village Resouce Center at Hadonahalli.

Table 18. Source for information on climate change

Computer	Internet	Govt.	Radio	TV	Mobile phone	Telephone	VSAT	CD/ DVD	Movie	Campaigns
0%	9%	38%	47%	100%	0%	0%	38%	0%	0%	0%

It can be concluded that all the farmers need to be knowledgeable about scientific weather forecast at hobli level from Television news, VSAT and radio programmes. The same information should be provided on their mobile phones/telephones. They should be taught to interpret the data sent by SMS or available on the internet web site of IMD, DMC and UAS.

Table 19. Frequency of use of ICT tools

ICTs used	Daily %	Weekly %	Monthly %	Rarely %	Never %
Cell phones	78%	0%	0%	16%	6%
Landline	31%	0%	0%	0%	69%
Computer without internet	6%	0%	0%	0%	94%
Computer with internet	6%	0%	0%	0%	94%
Video	31%	16%	0%	0%	53%
Satellite Video conferencing at Village resource centers or Krishi Vigyan Kendra	0%	34%	84%	0%	0%
Television	100%	0%	0%	0%	0%
Radio	0%	63%	0%	25%	0%
Blogs	0%	0%	0%	0%	100%
www.bangaloreimd.com	0%	0%	0%	0%	100%
Call 1800-180-1717 (IMD)	0%	0%	0%	0%	100%
www.dmc.kar.in	0%	0%	0%	0%	100%
Call 222 11 118 (IMD)	0%	0%	0%	0%	100%

Table 19 illustrate that 78% if the respondents use cell phones daily compared to 31% who use landlines. 31% use CD/DVD on their video players daily compared to 6% who use computers. However, 100% of the respondents view the television daily. Only about 16% use the CD/DVD weekly compared to 34% who visit the Village Resource Center. 63% of the respondents listen to the radio occasionally in the week, while 84% of the respondents visit the Krishi Vigyan Kendra once a month to gather information from digital sources about agriculture. About 16% rarely used and 6% never used mobiles as they were elders who shy away from use of these gadgets. The elders have access to the mobile phones in possession by their children.

It can be concluded that majority do not use the computers, internet, the call center nor the websites for accessing information on weather forecast. There is an decrease(69%) on the use of landline phones. But only 50% mentioned that they do not use the CD/DVDs.

Sci-panchang

The new model suggests that the exhaustive data available with the IMD and KSNDMC on rainfall, wind and temperature will have to be imprinted on the Krish-Panchang calendar in an understanding manner. Further, on an interactive mode of using the calendar would surely enhance the awareness of scientific weather forecasting.

The calendar can be designed with boxes the colours such as Green, Yellow and Red suggesting as safe, partially safe and unsafe can be displayed. Further, the farmer can tick mark either one of them for the next day as per the forecast given by the scientific community. This will be the change agent for farmers to differ from the traditional mode to the scientific mode of communication. The farmer can later append the calendar by ticking the colours again on the other side of the date in blank boxes of the three colours stating if the scientific forecast has benefitted him. The colour green would indicate has benefitted, yellow as partially benefitted and red as weather forecast not relevant.

IMD must introduce a brand name say 'Sci-Panchang' calendar and the frequency of the scientific weather forecast given by IMD, KSNDMC and allied bodies will have to be increased through their mass communication channels. The weather forecast should be significant at Hobli (Village) level and not district or taluk level.

Conclusion

Communication is a vital element in human society. The communication of scientific weather forecast to the farmers is a primary need. To conclude the majority of the farmers of Hadonahalli need scientific weather forecast with rainfall as a priority during the months July to February. However, they need the wind and temperature forecast only on extreme climatic conditions. The Rashtriya panchang and Krishi Panchang have value based information about astronomy based on India calendar era. These names of months and stars are well known by the old generation farmers and those youngsters who listened learnt the names. The elders always relate to the weather forecast referring to the names of the months and stars given in the regional panchang. Presently due to climate change phenomena universally, the tendency among farmers has changed in that they have to begin to follow the scientific weather forecast. Many have mentioned that electric power shut down is causing a problem to totally depend on electronic gadgets. However, the newspaper, mobile phones and landline phone are the only ones that are constantly available for continual communication. The farmers have a need to be communicated about rainfall forecast sometimes even 12 hours ahead. This will enable the farmer to take a decision thereby saving in cost of farming. The Scientific weather forecast is mainly required by the farmers during the North East (Hingaru Malai) season at hobli level. Farmers presently do not completely follow the weather forecast from the mass media as it reflects to the district level forecast and sometimes taluk level which have less significance to the hobli level farmers.

Though the Antariksha method of weather forecast as mentioned in the study of Varshneya MC *et al.* (2009)[24], is predicted well in advance and published in local dailies in Maharastra, this study proves that there are alternative scientific methods that people are following today among all demographics using Information and Communication Tools. The study finds that the effect on scientific weather forecast is rapidly increasing due to the exposure of ICT tools and accessibility to the farmers. If Hobli wise daily rainfall forecast are made more frequently, say every half hour during the cultivation to harvest seasons using the existing popular communication channels as Television, Radio, VSAT, CD/DVD, and Mobile phones then the farming community would by enlarge follow and on experience trust the scientific weather forecast in future. This model will ensure that farmers save a lot of expenses and avoid tendency of taking loans to meet farming expenses and later resulting in suicidal attempts due to inadequate repayment capacity from income arising from specific agricultural produce among food and cash crops. The effective outreach to the farmers need to begin with the interactive model namely the Sci-Panchang model suggested in this study.

References

1. United Nations Framework Convention on Climate Change (UNFCCC). Retrieved from http://www.itu.int/themes/climate/
2. Peter Newell and Matthew Paterson (2010), *"Climate Capitalism"*, Cambridge University Press, New York, p 37.
3. ibid. p. 156-157.
4. Mixed 2010 report card on anti-poverty goals highlights challenges ahead for September Summit. Retrieved from http://un.org//millenniumgoals/pdf/ MDG_A_ Main.pdf, June 23, 2010.
5. United Nations Millennium Development Goals, retrieved from http://www.wmo. int/pages/themes/weather/developmentgoals_en.html.

6. Mixed 2010 report card on anti-poverty goals highlights challenges ahead for September Summit. Retrieved from http://un.org//millenniumgoals/pdf/MDG_A_Main.pdf, June 23, 2010.
7. World Summit on the Information Society, Geneva 2003, Plan of Action, Tunis, 2005, 12 December 2003. Retrieved from http://www.itu.int/wsis/docs/geneva/official/ poa.html.
8. WCC-3, Working to make climate services available to all, Geneva, 31 August to 4 Sept 2009. Retrieved from http://www.wmo.int/pages/mediacentre/press_releases/pr_860_en.htmlhttp://www.wmo.int/pages/mediacentre/statann/statements_announcements_en.html ibid. p. 2.
9. Jayalakshmi Chittoor (2008), Climate Change: An overview-Shared responsibility for climate control retrieved from http://i4donline.net/May08/May08.pdf.
10. Dr Shahriar Hossain, Oct 3, 2009, Use of ICT for adaptation to climate change. Retrieved from http://www.thedailystar.net/newDesign/news-details.php?nid=107998.
11. Gokul Raj M P (2005), *"Role of Radio and newspapers in Dissemination of Agricultural Technology in Karnataka – A Comparative study"*, Department of Communication, Bangalore University, Bangalore.
12. Annual Report 2008, (2009), Indian Meterological Department, Ministry of Earth Sciences, Government of India, The Meterological Office press, Pune.
13. Director General of Meteorology(2009), "Rashtriya Panchang Saka Era 1932, Kali era 5110-11, 2010-2011 AD", The Controller of Publicaitons, Delhi. Pg iv to vii.
14. Activities and progress of the work, KSNDMC. Retrieved from http://www.dmc.kar.nic.in/Activities%20of%20DMC_2004-05.htm.
15. Shankar Bennur (2009), Project to install telemetric rain gauges nearing completion, The Hindu. Retrieved from http://www.hindu.com/2009/11/09/stories/2009110959680600.htm.
16. UN Secretary-General at ITU Telecom World 2009: Information Technologies Part of Climate Change Solution (2009). Retrieved from http://climate-l.org/news/un-secretary-general-at-itu-telecom-world-2009-information-technologies-part-of-climate-change-solution/.
17. Ramakrishnan T R (2010), *"The Use of Information and Communication Technology (ICT) for fostering rural prosperity and grassroot development in Karnataka"*, *Conference proceedings,* Knowledge Utsav, August 28th 2010, Jain Group of Institutions, Global campus, Bangalore.
18. Ramakrishnan T R (2009), *"Analysis of ICT enabled farmers using Village Resource Centers (VRC's) for farm technology transfer from Bengaluru"*, Conference proceedings, 18th AMIC annual conference July 13th -16th 2009, New Delhi.
19. Secretariat of Agriculture Department Annual Report 2008-2009, Government Press, Bangalore.
20. Sridhar V and Shamsher Singh, August 28, 2010, *"Rural India's communication divide"*, The Hindu, Bangalore.
21. Satish Kumar B S, August 11, 2010, *"When something is in the air"*, The Hindu, Bangalore.

22. DH news service, September 4, 2010, "5,220 GPs go online", Deccan Derald, Bangalore.

23. Vaidyanathan A, (2010),"*Agricultural Growth in India: Role of Technology, Incentives and Institutions*", Oxford University Press.

24. Varshneya M C, Nanaji Kale, Vaidya V B, Kane P V and Vyas Pandey (2009), *"Forecasting and validation of rainfall for Barshi in Maharastra based on astro-meterological principle of rainfall conception"*, Vol. 13, No. 3 Asian Agri-History, ICRISAT, Secunderabad, India.

Annexure 1 (Questionnaire)

Part 1

1	Name		Cell No		
2	What is the main crop cultivated?	Season 1	Season 2	Season 3	
3	Whose opinion do you follow for weather forecast? Please ✓	From general conversation			
		From past experience			
		From the Calendar			
		From self imagination			
4	Do you go by the opinion from the above on situations?	Depression (Ocean)	Ordinary day (due to heat)	Pocket area (Certain areas)	
	Please ✓	Y / N	Y / N	Y / N	

5	Which stage of farming due you give importance for weather forecast? ✓ **(Importance)**	**Rainfall**					**Wind**					**Temperature**				
	Not at all (1), Not very (2), No opinion either way (3), some-what (4), extremely (5)	5	4	3	2	1	5	4	3	2	1	5	4	3	2	1
	Preparing the land for sowing															
	Spacing & Sowing															
	Application of manure and fertilizers															
	Weed management															
	Water management															
	Pest management															
	Disease management															
	Harvesting															

6	What is the period of forecast expected by you?	24 hours	48 hours	3 days	5 days
	Please ✓				
7	Were you successful in your decision, based on your perception/intuition?	Always	Sometimes	Never	
8	During which season were your decision right?	Pre-monsoon	South-west monsoon	North-East monsoon	
	Please ✓				
9	During which stage of cultivation do you need proper weather forecast? **(Importance)**	**Rainfall**	**Wind**	**Temperature**	

	Not at all (1), Not very (2), No opinion either way (3), somewhat (4), extremely (5)	5	4	3	2	1	5	4	3	2	1	5	4	3	2	1
	Preparing the land for sowing															
	Spacing & Sowing															
	Application of manure and fertilizers															
	Weed management															
	Water management															
	Pest management															
	Disease management															
	Harvesting															

Part 2

1	Do you follow the mass media for scientific weather forecast? ✓	**Yes**	**No**

2	If YES which of the following media do you refer? ✓	**Newspaper**	**Radio**	**T.V**	**Mobile phone**	**Internet**	**Campaigns**	**Phone**	**Government office(VSAT)**

3	For which season do you follow the scientific weather forecast from mass media? ✓	Pre-monsoon	South-west Monsoon	North-East Monsoon
4	Do you take decision on the following farming activity from **the scientific weather forecast** from mass media?	**Rainfall** ✓	**Wind** ✓	**Temperature** ✓

	Always(A) Sometimes (S) Never(N)	**A**	**S**	**N**	**A**	**S**	**N**	**A**	**S**	**N**
5	Preparing the land for sowing									
	Spacing & Sowing									
	Application of manure and fertilizers									
	Weed management									
	Water management									
	Pest management									
	Disease management									
	Harvesting									

5	Do you trust the scientific weather forecast? ✓	**Always**	**Sometimes**	**Never**

6	Are you aware of the Climate Change? ✓	**Yes**	**No**
7	Have you felt the CHANGE in the CLIMATE in your practice in agriculture? ✓	**Yes**	**No**

8	If 'YES' do you feel that you should trust the Scientific Weather Forecast? ✓	**Yes**	**No**
9	In what way have you experienced the Climate change?		
10	**Are you aware that the Climate change is due to the following reasons?** ✓	**Yes**	**No**
	Change in Atmosphere (Air, water vapour, CO_2, clouds, particles)		
	Change in the Hydrosphere (continental ice-sheets, mountain glaciers, seas ice and surface snow cover)		
	External changes (Volcanic eruptions, Sun's output, Earth's orbital variations)		
	Changes in solid earth (continental drift, mountain building, erosion and siltation)		
	Large rises in sea levels		
	Massive changes in plant existence		
	Massive change in animal numbers		
	Massive change in Land-sea boarders		
	Change from treeless area to temperate forest		
	Change from temperate forest to deserts		
	Change in temperature (Extreme temperature/climate)		
	Change in rainfall (Floods, less rainfall during rainy seasons)		
	Change in wind		
	Change in Draughts		
	Change in human population		
	Change in irrigation water supply		
11	**How did you know about climate change information?** ✓		

	Computer	Internet	Government office	Radio	TV	Mobile phone	Telephone	VSAT	CD/DVD	Movie	Campaigns

12. How often do you use the following ICT tools ✓

ICTs used	Daily	Weekly	Monthly	Rarely	Never
Cell phones					
Landline					
Computer without internet					
Computer with internet					
Video					

Satellite Video conferencing at Village resource centers or Krishi Vigyan Kendra					
Television					
Radio					
Blogs					
www.bangaloreimd.com					
Call 1800-180-1717 (IMD)					
www.dmc.kar.in					
Call 222 11 118 (IMD)					

Conservation and reuse of resources at household level to mitigate the effects of global warming

S. Muraleedharan

Department of Economics, Maharaja's College, Ernakulam, Kochi, India.
[E-mail: muraleedharanvarsha@yahoo.co.in]

Introduction

'Charity begins at home' was a well recognised saying for a pretty long time. It may be reshaped for the contemporary world as follows: 'Conservation begins at home'. Most often, it is argued that mitigating the negative externalities of global warming lies with the international community, national, provincial and local government, civil society at large etc. The specific role of the individuals is not properly and seriously deliberated upon.

The present paper is an attempt to link the day to day household activities and global warming. It must come, through conservation and reuse of resources within the household premises. It can be mainly done in the case of energy, water, and paper and kitchen waste. Any conservation or reuse of these resources, in fact, leads to household level effects to mitigate global warming. The study tries to quantify and express that value in money terms.

There are three sections in this paper, apart from the introduction. Section I refers to the general discussion on the resource conservation at the household level. The second section which is the focus of the paper, narrates the household experience of the author in resource conservation for one year. The emerging conclusion are summarised in the last section.

Discussion

Fresh water is not infinite but essential for sustaining life. Diversified use of water sets an atmosphere of competition to grab water from various sources. In the context of vulnerable supply of water and rising demand for it, water conservation becomes part of top priority even at house hold level. This is very significant in the backdrop of excessive waste in the use of water. There is tremendous scope for water saving at the domestic level. Political system is effective to provide water among the common man in the third world[1]. This can affect human health and welfare. The real challenge is to develop and apply water saving technology and management methods[2].

Water and energy conservation is fundamental to the resource management at household level. Water use at household level is normally categorised into 8 purposes viz. shower, clothes, toilets, dish, baths, leaks, faucets and others[3]. The share of water use for showers, clothes, toilets and faucets came around 20 per cent each in American Society. The relevant data is represented in Table 1.

Table 1. Use-wise water consumption at household level

Purpose	Quantity in gallons	Percentage
Showers	8.8	19.5
Clothes	10.0	22.17
Toilets	8.2	18.0
Dish	0.7	1.5
Baths	1.2	2.7
Leaks	4.0	8.8
Faucets	10.8	23.9
Others	1.6	3.4

Source: Handbook of water use and conservation, American Water Works Association 2010

Leakages do not have an insignificant role in water use (8.8%). The average water consumption of a person is 3000 gallons per day and a four member family requires 12,000 gallons for bathing, cooking, washing, recreation and watering[4]. Outside watering requires 5 to 10 (37.75 litres) gallons per minute. Five to seven gallons of water per flush are consumed in toilets. Another study shows that average consumption of water at household level is 50 gallons and its 65 percent is used for bathroom. The laundry and kitchen purposes constitute 25 percent[5] while another 10 percent is meant for outdoor uses. One study by the Agricultural Research and Co-Operative Extension, Penn state regarding household water conservation (2009) points out that the major items of domestic water use are toilet, clothes, shower and faucets. The average consumption is 69.3 gallon per person per day in US. The study reveals that 4 gallons per minute can be saved by turning off tap while brushing the teeth. It totals to 200 gallons a week for a family of four[6]

Lighting is a major power consumption item at household level. The basic necessity is that lights must be turned off when it is not necessary. The standard version is that the use of CFL (Compact Fluorescent Lamp) can reduce power consumption by five times[7]. Another study reveals that energy saving in terms of CFL is 50 per cent[8]. Electrical items shall be bought after verifying the energy rating and replace all the aging equipment with new ones[9]. In fact what are required are efficient lights which arc also environment friendly[10].

Another, resource conservation is possible in the case of paper at the household level. It would be interesting to link the paper with trees. It is difficult to measure exactly the number of paper sheets from a tree. It depends on the type and size of the trees. Calculation shows that 4 feet 4 feet by 8 feet have 128 cubic feet of wood. It can bring 90,000 sheets of bond quantity paper[11]. Related information is that one ream paper (500 sheets) consumes six per cent of an average tree. In other words a ton of printing and writing paper turns from 24 trees (one tree has the size of 400 feet and six to eight inches in diameter)[12].

The household experience

Water conservation

Attitude in the kitchen is very significant in saving drinking water. A lot of water can be conserved if washing of vessels and utensils are undertaken with practical wisdom. In the rural areas in the past, grand mothers used to wash small plates by collecting water in a large vessel and initially keeping medium and small ones in it. As they had to carry water from a distant pond or well, it was necessary to cultivate such a practice. However in the urban area water is available at the kitchen slab except in the case of very poor people. As there is no physical strain, individuals wash all vessels/utensils in running water. A fully opened tap can facilitate 2 litres of water per minute. If a person does the washing of plates and vessels in flowing water there would be overuse of water. How much of water is overused due to this kind of household behaviour? The author observed the level of water saving (avoiding of excess use) for one year and the obtained result is reported in Table 2.

Table 2. The degree of water saving through separation of cleaning & washing of vessels & utensils for one year

Item	Large	Medium	Small	Cleaning in mts.	Washing in mts.
No./ Time	135	255	319	598	133
Average time				0.843	0.188
Without large	-	42	54	40	14
Average time				0.417	0.146

Source: Documented data for one year (2009-10) during self use

Vessels and utensils are divided into large, medium and small groups. This categorization is based on the size and shape of each item. Certain vessels have many curves and complicated structure. It takes more time and hence it can be treated as large or medium. That shape is a criteria rather than the size in classifying an item as large even if it is small in size. The data shows the number of the various vessels washed and the time taken to clean and wash the items. The average time taken is 0.843 and 0.188 minutes respectively for cleaning and washing. The average time reduces to 0.417 and 0.146 minutes once activity takes place without large vessels and utensils. The total time for cleaning and washing is 12 hrs and 18 minutes and the cleaning time is 82 per cent of the former. In other words, 82 percent of water consumption is saved if cleaning and washing are done separately.

It can be noted that the clearing time for large items is 1.89 and 2.36 times more than that of medium and small items respectively (Table 3). Almost the same difference exists for medium and small items in the case of actual wash. It seems that the practical wisdom is gradually disappearing with urbanization and the availability of water at the sink of the washing area of the kitchen. The focus interview made with a few household wives reveals the following details.

- Cleaning and washing are done separately by old people, especially in the rural area.
- Separation comes down if a household wife has the habit of washing most of the items then and there.
- It is very difficult to execute the 'separation principle' if the household servant is dominating in the kitchen.
- Majority accepts the 'separation principle' but reluctance in implementation.
- Some of them wash speedily while cleaning in running water.

Table 3. The degree of water saving through separation of cleaning & washing of vessels & utensils for one year (size-wise)

Item	Large	Medium	Small
Average cleaning time (mts)	4.430	2.345	1.875
Average washing Time (mts)	0.985	0.521	0.416
Average cleaning time without large (mts)	-	0.952	0.741
Average washing time without large (mts)	-	0.333	0.259

Source: Documented data for one year (2009-10) during self use

One of the major water issues in Ernakulam area which is in the coastal belt is that water is available at every household premises but most of the people rely on drinking water for daily purposes. The use of drinking water ranges from cooking to dog bathing and watering the garden. The main reason for the non use of nearby well water is the colour of the water. It contains iron compounds so that it is difficult to use for daily purpose. However, a little arrangement and modification can enable the use of well water in a better way. Firstly the water can be pumped to an overhead tank of 1000 litres capacity on the previous day. The coloured substances in most cases will settle at the bottom of the tank so that the water can be used for various purposes like cloth and car washing, toilet, gardening etc. It is a criminal waste to use drinking water for these purposes. It needs to be expressed in monetary terms to explain the gravity of the situation. Flushing of water in the toilet once requires 10 litres of water. Assuming a person uses the toilet five times a day, a four member family requires 25 litres of water per day. Considering the time taken at the working place and educational institutions, 15 litres (5×3) per day can be taken as an average quantity. If so, the total annual quantity is 21900 litres Table 4. Every week, five buckets of water ($10 \times 5 = 50$ litres) are used to wash the car, which amounts to 2600 litres

(50 × 52) annually. 25 garden pots require 25 litres (25 × 1) per day and the total quantity required water comes to (25 × 240 = 6000 litres). Four months are deleted to calculate the annual quantity in order to adjust for the rainy periods. Moreover, average watering quantity is taken after considering the moderate temperature in September and also avoiding the extreme quantities during winter and summer seasons. The 40 litres of water are necessary to wash the cloth on every day. As a result, the total quantity for cloth washing comes to 14600 litres (40 × 365 = 14600). Thus the total annual quantity of well water use is 45100 litres in terms of toilet, gardening, washing clothes and car (Table 4). If drinking water is used, the required cost at current market rate is Rs. 334.

Table 4. Extent of drinking water saving

Purpose	Average Use in litres /day	Annual Quantity in litres.
Toilet	60	21900
Car wash	50/ week	2600
Garden pots	1/ pot	25pots*280 days = 6000
Cloth Washing	40	14600
Total		45100

Source: Documented data for one year (2009-10) during self use

How to reduce the use of LPG

It is a part of modern life style that only boiled water is consumed which is bacteria free. It is so deeply rooted in the mind of every one. It is also clear that boiled water contains less oxygen and in this sense it has less quality. A five member family normally consumes about 10 litres (5*2) of water that may require 30 minutes of boiling every day. Which means that two cylinder of LPG (one household LPG cylinder in India can normally use for 95 hours), is necessary for one year. This can be avoided if filter is used, as this method does not require any power and gas consumption.

Sorting of kitchen waste is another way of reducing the use of LPG. If bio degradable waste is differentiated from non degradable items, a small bio plant will be helpful to generate gas for cooking. After keeping degradable bio waste with kitchen generated waste water for one day the cooking gas for 30 to 35 minutes is got. Like water filtering this can also save one cylinder of gas every year. It implies that water filtering and using a bio gas plant by a nuclear family can reduce mainstream LPG consumption by 4 cylinders every year. Considering the current cost of LPG used for domestic consumption, the cost comes to about Rs. 1,400.

Three case studies were made to understand the method of household waste in three locations in the coastal belt of Ernakulam District.

Panampilly Nagar (PN) and Gandhi Nagar (GN) are two adjacent locations in the heart of Kochi city. The former consists mainly of high income families while the latter is dominated by middle and low income categories. A random survey conducted in PN in 2006 showed that 75 per cent of household waste consisted of bio degradable materials and 88 per cent of family on average generated one or two 'kit' of waste per day (one kit can carry a waste of 5 kg). 60 per cent of the families made use of the formal collection of system and a quarter of them had the habit of burning the waste. Though 75 per cent of the waste was biodegradable, only 3 per cent families were using it for compost. The behaviour of burning the house hold waste (25 per cent) needs to be discouraged.

Table 5. Method of disposal of waste (in %)

Method	PN	GN	EDY	Kulgi
Waste Bin	18.33	33.3	24	
Door to door collection	41.67	18.3	48	
Burning	25	36.6		
Compost	3.33	8.3		
Throwing away	10			
Own ground	1.67		3	5
Biogas/compost			25	95

Source: survey data; Note: each case consists of a sample size of 60

Household behaviour in Gandhi Nagar (GN), where relatively low income groups are residing, is different from PN to a certain extent. Door to door collection was more in PN while the use of waste bin (33.3%) and burning (36.6%) were more in GN (Table 5). Compost making was more in GN than in PN. The surveys in Edappally (EDY) and Kumbalangi (KULGI) were conducted in 2009 where the former is in the suburb of Kochi while Kumbalangi is India's first 'tourism village' near Kochi City. The study in Edappally was conducted with the help of a residential association called "SWAREKSHA". The behaviour of burning the waste seems to be in the reverse gear in recent days. More than 70 per cent of the households make use of waste bin (24%) and door to door collection (48%) for disposing household waste. Conversion of house hold waste to compost or biogas is more here (25%) but very much high in Kumbalangi (95%). The high percentage of biogas plant in Kumbalangi is the result of state intervention in the context of promotion of rural tourism. More over the Edappally data reveals that public interference can reduce the habit of burning waste and enhance the trend in processing household waste (compost/biogas). The awareness programme and collection system by local body (Kochi Municipal Corporation) and residential association were responsible for better disposal methods of household waste.

LPG consumption can be reduced by way of tapping solar energy as well. The demand for hot water increases with change in seasons or the presence of aged people. If pure Ayurvedic principle is followed hot water must not normally be mixed with ordinary water, especially in the case of aged persons. If a person uses two buckets of water for a bath, it requires 20 litres of hot water. Practical experience shows that heating 20 litres of water takes one hour. That is, two hours are necessary to get 40 litres of water for two aged persons. This means that 730 hrs of LPG consumption is necessary for a year. This is equal to the use of 7.3 LPG cylinders per year. This is equal to Rs. 2657.20. Except three months (rainy and winter), solar energy is trappable in a place like Ernakulam with an investment of Rs 35,000. If aged persons or demand for hot water for bathing are not in a house, the hot water can be used for cooking or washing purposes.

Reuse of the paper

Paper is an integral part of the modern life, unlike many other resources paper can be reused in multiple times. That is, the same paper can be re-used at different stages. Used paper may be categorised into the 'sheets and pieces'. Personal experience of the author shows that used full scap papers are available in the contexts of Photostat, letters, receipts, accounts, cover etc. One can list a lot of them. Observation by the author for a year (2009) shows that the author could save 20, one sided sheets which was later used for various writing purposes. This is equal to 10 full scap papers. Pieces of paper are available at various occasions like small receipts, files, unwritten or no printed portion of the used papers, small notices, and note books and so on. The collected pieces of paper are spread on a foolscap paper to determine the number of pages it can make. Thus, 70 one sided sheets were acquired in 2009 and it has reached 37 one sided sheets till 31st August 2010. Thus the average foolscap paper in terms of pieces of paper comes to about 35 full sheets in a year. These reused papers will again be stored in a packet so that full sheets can be used for packing, cleaning, compost, etc. Thus a well planned used paper collection leads to multiple uses and a 45 full sheet paper (35 + 10). This is again used as waste paper in daily life, which is equal to 90 full sheets. More than 100 sheets are directly used as waste paper as there is no chance for writing. Thus the total annual saving of the paper due to multiple uses is equal to 200 sheets and it is equal to 3/125th (0.024) of a tree considering one ream is equal to 6 per cent of a tree. In short, every sheet of paper saved is same as preserving 0.012 per cent of an average tree. Thus the current cost of the 200 sheets is equal to Rs 200 per year.

Conservation of household energy

Household Energy saving is equal to the power generation as advertised by the public relations department of Kerala State Electricity Board. One immediate action for household energy management is to replace the fancy shades or decide not to use it. Otherwise every fancy shade requires bulbs of higher watts in order to get sufficient light. That is, a bulb of 60 or 100 watts have to be used for sufficient light. This enhances power consumption by 50 or 150 per cent. If electronic ballasts are used instead of ordinary chalk, power consumption can reduced by 10 per cent. What is required is to replace incandescent lamps and tube lights with CFL. Power consumption can be saved by pursuing different strategies. In bath rooms, outer walls and works areas, 5 watt CFL can supply light worth 40 watts of an incandescent lamp. If a 100 watts lamp is live for 10 hours, the meter registers one unit. That is a 40 watts ordinary bulb registers one unit if it emits light for 25 hours. It takes 200 hours to become one unit in the case of a 5 watt CFL. Two bathrooms, one work area and three outer wall lamps of a house need only five watts CFL. Six 40 watt incandescent lamps are equal to 240 watts, which would take only 4.17 hours to account a unit. Meanwhile six five watts CFLs would take 33.33 hours to reach the same level (i.e. one unit). Two living rooms, dining hall, normal dressing room, kitchen and car porch of a house requires six 15 watts CFLs. If we follow the logic just explained with the additional condition that these rooms have 60 watts incandescent lamps, the resultant power saving would be 3.65 units (Table 6). Suppose of the rooms is used for study purposes it is better to have 20 watts CFL, as the power saving will marginally reduce to 3 units. If a house has a stair case which is dark but is close to a door opening outside, energy can be saved easily by keeping the door partially open during day time (of Course, a mosquito net door is necessary), thus power consumption of a 15 watts lamp can be avoided for about 10 hours. There is a saving of 150 watts per day, which is equal to 15 per cent of a unit.

Table 6. Power saving in the household

Type	Watt	No.	Total Watts	Hrs for 1 Unit	Power Saving in units
CL	40	6	240	4.17	
CFL	5	6	30	33.33	8
CL	60	6	360	2.77	
CFL	15	6	90	10.11	3.65
Fridge			3 hrs. cut off		0.39
Total					12.04

Source: Household Practice

In the case of a fridge, power consumption can be minimised if it is switched off for three hours per day 1.04 units per day after 8'O clock in the night. If the average consumption of power per day by a standard fridge in the house is 1.04 units per day (380/365 days), a three hour power cut to fridge implies a saving of 13 per cent of a unit, which implies that approximately a unit is saved in every '8 days'. In other words the saving is 4 units per month. Thus the total saving in every 10 hours of all the cases explained above is equal to 12 units (Table 6).

Conclusion

Household behaviour is very important in reducing the impact of global warming. Prudence and practical wisdom are pertinent mantra to achieve the same. This paper accounts a few household level cases, which are indirectly helpful in mitigating global warming.

Many methods can be adopted to reduce LPG consumption. Household experiments show that LPG consumption can be cut down by four domestic cylinders yearly by sorting of waste and setting up of biogas plant and, using non-electrical water filters. Four household level surveys reveal that burning and throwing of household waste will come down with offering processing methods like composting, biogas plants etc. Successful operation depends on governmental and other public interventions.

Multiple use of waste paper based on size-wise sorting and purpose oriented writing can put down the additional demand for fresh sheets of paper. Household documentation tells that more than 200 sheets of paper can be saved every year through sorting and purpose oriented use of waste paper. This is equal to 2.4 per cent of a standard sized tree and similar actions by millions of citizens can slow down the process of deforestation, which is very much relevant to global warming. The money cost for 200 sheets is Rs. 200.

Water conservation through household behaviour is pertinent for both resource management and global warming. Water use in cleaning and washing of vessels and utensils is a contextual point. The average time taken to clean and wash vessels and utensils is 0.843 and 0.188 minutes respectively. The water saving due to the separation of cleaning from washing is 82 per cent of the total time taken for both cleaning and washing. If the large vessels are excluded, water saving is 74 per cent of the total time required for cleaning and washing. That is, water use will increase by 8 per cent if running water is used for both cleaning and washing. When large, medium and small vessels and utensils are taken together, the average time changes with the size of the vessels and utensils. Well water use reduces the demand for drinking water by 45100 litres so that it can save Rs. 334.

Practical wisdom can lead to setting up of power houses within houses. If all the fancy shades are removed or they are not in use, the power consumption comes down by 50 or 150 per cent per point. A further 10 per cent can be saved if electronic ballasts are used for tube lights. If six each 5 and 15 watts CFLs are used in the place of six 40 and 60 candescend lamps respectively in a standard house of (1000 to 1200 sq. ft.), a household can save 8 units of energy for every 10 hours. This is equal to power generation 'within house'. A three hours cut off power to fridge every night (especially between 6.30 and 10.30 pm) leads to an additional power saving of 4 units every month or about one unit for 10 hours. Thus total gain reaches to 12 units.

Remember when we use electricity, water, paper and burn things we are leaving a carbon foot print and the more we use the bigger is our carbon foot print. This carbon foot print effects global warming and climate change, so be cautious while using them.

Acknowledgement

Authour is grateful to Prof. Mary Joseph, Dr. E M Mohammed, Dr. N. Shaji and Nirmala G Pillai for their valuable comments.

References

1. Eckholm, E P (1982) Down to Earth, Environment and Human Needs, Affiliated East-west Press Pvt. Ltd., New Delhi.
2. Bucks D A, Sammis T W and Dickey G L (1990) Irrigation for Arid Areas. In: Hoffman, GJ *et al.* (eds), Management of Farm Irrigation Systems, ASAE monograph, St. Joseph, mi pp. 513-542.
3. www. Drink tap .org.
4. www. City of Tulsa.org.
5. Keller Stefanie's (2010) : Optimisation at House, Sustainable Sanitation and Water Management, www. sswm.info.
6. www. Household Water Conservation, water use it wisely.com.
7. --------(2010): Buy Products That Make A Difference, US Enviornmental Protection Agency, Us Department of Energy.
8. Linderman, Peter (2010): Save- in- House Energy.com.
9. Trembly Jr K R (2009): Lighting, www. Cext.colostate.edu.
10. Ravindran M (2008) : Energy From the Environment, Paper presented in the Seminar conducted by ABS, British Council & Cusat, Kochi.
11. www. Infoplease.com, 2010.
12. www. Green @work magazine, 2000-09.

Domestic chullah and climate vulnerability — A perceptible measure to combat carbon dioxide emission from domestic fire wood use

Sreedharan K.

Integrated Rrural Technology Centre, Mundur, Palakkad, India
[E-mail: irtcpalakkad@gmail.com]

Introduction

Global warming and consequent effects of climate change, sea level rise, agricultural lose and umpteen other consequences have been a subject of global concern. Global increase in the use of fossil fuel which releases larger and larger amount of CO_2, combined with increasing deforestation and tree cover loss which reduces the carbon sink causes steady increase in the CO_2 level in the atmosphere. From a mere 240 ppm the atmospheric CO_2 level has increased to about 370 ppm during the 20th century. It is feared that with in few decades the level may even cross 550 pm which is highly disastrous. The disaster will be unbearable to coastal countries and island nations. It is estimated that before the end of the present century the sea level may rise to a dangerous level if corrective measures are not taken. India having 7000 km of coastal stretch and many island formations will be one of the worst affected countries. Global as well as local efforts are essential, that too at the earliest possible time to mitigate CO_2 level rise. Every unit of heat generated by burning of fuel weather it is coal or firewood, petrol or diesel or kerosene has its on contribution to atmospheric CO_2.

Cooking is a high energy consumption activity causing substantial CO_2 emission and natural resource exploitation. The conventional chullahs used in common man's kitchen consumes about 5-10 kg of fire wood every day. It may vary depending on their lifestyle and economic capability (The cooking gas and the electric stoves of middle class and upper classes will cause its own thrust on the environment). Majority of households in rural India is using the conventional three stone chullah and it is high time that they be replaced by high efficient and less firewood consuming modern type of chullahs in order to save the environment, not to speak of its economic and health aspects.

Materials and methods

The ordinary firewood is a carbon rich organic material. It contains about 50% in weight by carbon. This means burning of 1 kg of fire wood will release about half a kilogram of carbon. Each carbon molecule (of molecular mass 12) will be combined with two oxygen molecules each of mass 16. The half a kilogram of carbon in 1 kg of firewood will require about 1.2 kg of oxygen, there by releasing 1.7 kg of CO_2. Thus every 1 kg of fire wood on burning will be releasing 1.7 kg of CO_2.

The efficiency of the conventional three stone chullah is less than 10%. This means about 90% of the heat is released in to the atmosphere as waste (Table 1). 1 kg of firewood will be releasing about 4000 kilo calories of heat on burning. Out of this, about 3200 kilo calories go wasted. This again adds to the atmospheric heat. Thus we see that 1kg of firewood on burning will release 3200 kilo calories of wasted heat and 1.7 kg of CO_2 (Table 2). Both of these will add up to global warming which the whole world is seriously worried about. Out of the 60 lakh households in Kerala, about 50 % are still using the conventional low efficiency chullahs.

The threat on the tree cover and forest for the fire wood need is another serious problem which also needs our attention. The money spent for firewood by the poor people is commensurate with the money needed for the daily bread.

Table 1. Generation of waste heat

Use of firewood per day per house hold	=	10kg
Heat produced per kg of firewood	=	4000 k.cal
Efficiency of conventional chullah	=	10%
Waste heat released per day per house	=	3200 k.cal
Waste heat released from 30 lakh house holds per year $30 \times 10^5 \times 365 \times 3200$	=	3×10^{12} k.cal
	=	3 lakh crore k.cal

Table 2. Generation of CO_2

Use of firewood per day per household	=	10 kg
Release of CO_2 by burning one kg of Fire wood	=	1.7 kg
Release of CO_2 per house per day	=	17 kg
Release of CO_2 from 30 lakh household in one year $30 x 10^5 x 17 x 365$	=	18.6 million tones

Considering all these aspects, the KSSP and its technical arm IRTC has conducted a serious research on conventional chullahs and energy conservation techniques and came up with two technical solutions to reduce the gigantic contribution to the global warming.

Results and discussion

1. Parishad aduppu (High efficiency chullah)

The ordinary aduppu (chullah) used in Kerala is built with three stones. It is wide open and its efficiency is only 10%. The improved parishad chullah with two main ovens and an ancillary one is provided with an upper hole for inserting fire wood, a lower hole for admitting air for combustion and a chimney for the exit of smoke and other combustion products. The different parts of the chullah, except the chimney, are made out of clay by potters. Special care is taken to optimize the dimensions of the fire hole, air hole and the connecting pipes to the ovens as well as to the chimney. The angle at which connecting pipes are fitted and the height of the baffle in the ancillary oven are regulated strictly to get maximum efficiency. The different parts of chullah are packed with a mixture of mud with high content of clay, rice husk and sand. The refractory nature of this combination prevents the loss of heat and enhances the efficiency of the chullah.

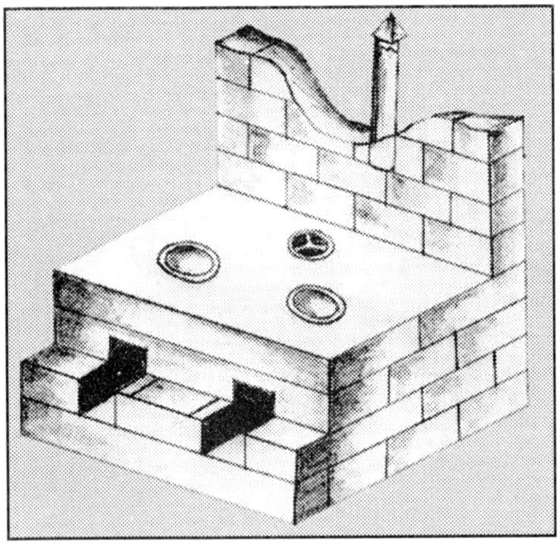

The efficiency of Parishad aduppu is about 30%. This is three times that of the ordinary chullah. This means the fuel consumption in Parishad aduppu is reduced nearly to one third compared to the ordinary chullah. The social and environmental impact of this is tremendous (Table 3).

Table 3. Reduction in CO_2 release using smokeless chullah

Use of firewood per day per house hold	=	10 kg
Efficiency of smokeless chullah	=	30%
Reduction in firewood use per day per house	=	6 kg
Reduction in CO_2 release due to this 6x 1.7 kg	=	10.2 kg
Reduction in CO_2 release per year if all the 30 lakh household replace their conventional chullah	=	$10.2 x 30 x 10^5 x 365$ kg
	=	11 million tones

Eleven million tones of CO_2 emission can be prevented by making use of this simple device.

Besides saving energy and fuel, Parishad Aduppu known as smokeless chullah, makes the kitchen free from smoke as the chimney draws out the vapour and smoke from the stove. The kitchen smoke contains hazardous and poisonous gases like carbon monoxide and benzo pyrene, and people, mainly women, who work in kitchen inhale such gases which are injurious to health.

IRTC has also developed community chullah which is found useful in hotels, schools and hostels. The efficiency of such chullah is very high, viz. 40-50%. Community chullah has to be built using brick and mortar depending on the size of vessels required for preparing food.

2. Choodarappetty (Hot box)

For a long time now, the people of Kerala have been using a thermal energy conservation device called kachippetty. It is a wooden box, lined inside with paddy straw. Rice and water in a vessel with lid just after boiling is placed in the kachippetti and it is covered with a wooden sheet. After some time rice gets cooked and meal is ready. The fuel required to cook rice if it was not placed in kachippetty is now saved. We often say that Science and Technology is developed as a part and parcel of day to day life and the use of kachippetty is the best example for this. Although the scientific principle behind this was not known to our ancestors, they were using kachippetty as an energy saving device. Now we know the principle that the total heat absorbed by the vessel at the time of initial boiling of the rice with water is sufficient to cook rice if this heat is not lost due to conduction, convection, and radiation. The kachippetty wooden box with inner lining of paddy straw, functions as a good insulator avoiding loss of heat.

Time has changed and in the modern era the kachippetty is a misfit in our kitchen. Moreover, the rice water may spill into the straw, and fungus and insects may grow in it. It is difficult to keep the kachippetty neat and tidy. The answer to these problems is the Hot box, devised and developed by IRTC, the research wing of KSSP. Hot box serves the purpose of kachippetty and it is made of thermocole which is a good insulator.

Using thermocole sheets square boxes were made and its efficiency to retain the temperature of water boiled for a few minutes was tested. The results were amazing and the cooling curve of temperature against time of boiled water placed in the box showed only nearly 10°C drop in 1 hour and it revealed that such boxes could be used as good insulators. Experiments were is rice cooking, the cooking vessel was removed from the stove and covered with the thermocole box and it was found that the rice got cooked in 40-50 minutes.

Cooling curve of a hotbox

Hot box, a cylindrical box of 30 cm diameter and 30cm height, is provided with an air tight lid. It is made of food grade quality thermocole using a mould where thermocole granules are subjected to hydraulic pressing and the product is injected out from the mould using steam.

Saving of LPG or any other fuel when a hotbox is used depends on how much it is used. A saving of 20% of LPG or kerosene is easily possible.

There are other advantages too. The stove will be free when the vessel is removed to the Hot Box and can be used for other purpose. Constant attention is required when rice containing vessel is on the stove, especially after boiling and now that burden is released. The cooked hot food or any other item can be kept inside the hot box. Since the cooling rate is slow even after 10-12 hours the food kept in Hot Box may be sufficiently hot for use.

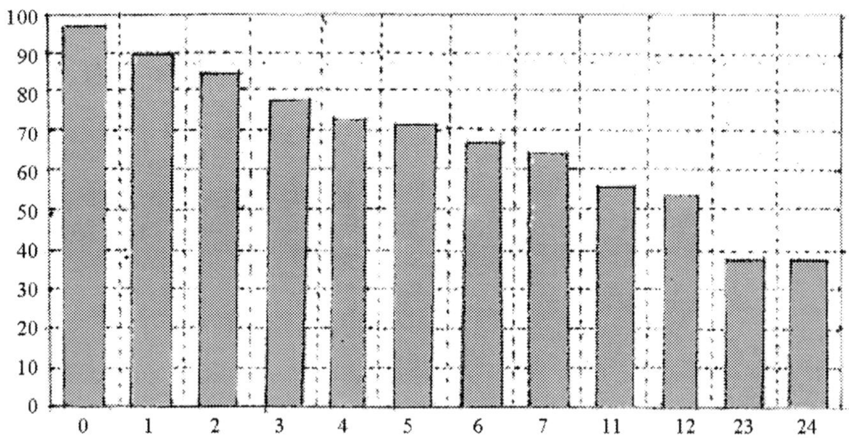

For cooking dhal, Bengal gram, etc hot box can be used. But in these cases the initial boiling should be to a larger extent.

Since it is a good insulator, hot box finds a larger number of uses like ice box for keeping fish, meat etc. a safe place for audio/video cassettes and CDs from the attack of fungi.

Subject Index